ROCKS AND MINERALS

OCTOPUS BOOKS

Acknowledgements

Most of the illustrations in this book show specimens from the author's collection, and were specially photographed by Mr Albert Payne of New Forest Printing. In addition, the author and publishers would like to thank the following agencies for permission to use their photographs: Aerofilms Ltd: 2, 3, 4, 5, 6, 8, 9, 20, 28, 42, 43, 62, 63, 70, 71, 77, 78; California State Library: 84; Camera Press: 84, 86; W. Churcher: 95; Fred Combs: 86; Ewing Galloway (by Burton Holmes): 3; Ewing Galloway: 4, 5, 70, 71, 78; Paul Laib: 74.

Preceding page
The concentrically-banded phosphate of aluminium variscite. Utah, United States.

This edition first published 1972 by
OCTOPUS BOOKS LIMITED
59, Grosvenor Street, London W.1.
ISBN 7064 0035 6
© 1969 by George Weidenfeld and Nicolson Ltd
PRODUCED BY MANDARIN PUBLISHERS LIMITED AND PRINTED IN HONG KONG

Rosemary Ransome-Wallis

Jan 1974.

Contents

1 (*left*) Lava erupting in crater. Etna 1971.

2 The Lot River Valley, east of Cahors, France, carved in clearly stratified limestone; it shows the erosive power of water on the easily dissolved limestone.

MEN HAVE always collected rocks and minerals for their intrinsic beauty and ornamental qualities, and the utility of natural materials has been the foundation of our development and prosperity.

Basically, rocks are aggregations of solid matter composed of one or more minerals, forming the earth's crust. They are commonly divided, according to their origin, into three major classes – sedimentary, igneous and metamorphic – which we shall deal with in more detail later. Some are obviously granular (e.g. granite, diorite, gabbro), while others are composed of grains visible only under a microscope. Rocks lack any definite structure, unlike their basic components, minerals. Minerals have a homogeneous chemical composition, unchanging physical properties, and typically, a definite crystalline form. A few minerals, such as gold, silver, lead and copper, are elements, but the vast majority are chemical compounds. Minerals may, and often do, occur in the form of crystals. Through the stone, bronze and iron ages the search for raw materials must have been of increasing importance, and in the middle ages landowners would employ prospectors to explore every crack and crevice of their estates for substances in demand. Yet the remarkable stories that can be read from the rocks began to be told only in the last three centuries. There is unmistakable evidence of great changes in climate and repeated inundations of the land by the sea. Violent volcanicity in areas now stable and quiescent, the growth and decay of great mountain chains, the dismemberment of continents and the scattering of fragments, and the growth of new land and the foundering of old can all be inferred from the observation of rocks. The signs and remains of extinct plants and animals give information on the changes that have led to modern species, and the peculiar distribution of the present flora and fauna can be understood only by reference to geological history.

Many of the discoveries upon which these geological concepts are based were made by doctors, clergymen, engineers and other amateur naturalists, and the field remains one of the few in which the amateur is able to contribute seriously to progress. He may share with the professional the heightened appreciation of land form and environment, and there is always a chance that he may find a yet unknown species of mineral or extinct animal. The quarry is not elusive or restricted to inaccessible places, but may be found anywhere and everywhere. Excavations for new buildings may be as fertile a source as cliff or mountain.

3 A bending road in the 'badlands' of South Dakota, one of the scenic wonders of the north-western United States. The horizontally stratified sediments are destroyed by periodic torrential rain and wind abrasion.

Many geological associations founded over the last eighty years have brought together teachers and amateurs, and their publications provide useful guides to accessible localities where rocks, minerals and fossils may be examined and collected. Most National Geological Surveys also produce maps and regional guides. The equipment required for sample-collecting is minimal. A hammer of about a pound in weight will suffice for dislodging suitable pieces of most rocks, and a chisel is often of value in extricating fossils. A hand lens with a tenfold magnification, specimen bags, a notebook and appropriate maps and guides are also desirable. Undoubtedly samples taken from their natural place are the most satisfactory, but often larger, cleaner and more spectacular specimens can be obtained from old mines and quarries. Samples can often be identified by reference to published work on the area and, failing that, most museums provide a service to the amateur and hold collections of samples for comparison.

The growth of a science inevitably stems from men of vision and persistence, and of these the earth sciences have had their share. Leonardo da Vinci recognised sea-shells in the rocks of high mountains and concluded that water had played a prime role in the formation of the land. He pointed to the frequent occurrence of layering in rocks and made sketches showing strata merely inches thick yet extending laterally for great distances. Less well known but certainly important to our knowledge of rocks was the work of James Hutton in the late eighteenth century. He realized that the processes taking place today at the earth's surface

4 Looking through the steep walls of the Rita De Frijole Canyon in New Mexico. The gradual uplift of the land allows the stream to cut an ever deepening gorge through coarse bedded sediments.

had always been happening, that the present was the key to understanding the past. The land is continuously being worn away by the pounding of the oceans, by running water, by heat and cold, by wind and by the polishing action of glaciers. Degraded rock waste so produced eventually finds its way to the sea to build beaches, sandbanks and deltas—new rocks in the making. These accumulations of sediment are usually layered or bedded and each stratum may entomb shells and other marine creatures, creating embryonic fossils.

William Smith, a canal engineer of the early nineteenth century, examined various types of rock and came to realize that he could identify individual strata by the fossils they contained and that younger layers always rested on older. These ideas laid the foundation of modern geology and have been refined and elaborated in extensive exploration and research. Mapping, collecting and interpreting in ever-increasing detail has led to the recognition of a sequence of strata which may total hundreds of thousands of feet in thickness. These have been allocated to several systems, each representing a period of time in the evolution of the earth and each readily identified by its fossils. There is no one place where this stratigraphical column is completely displayed and the whole can be pieced together only by collating information from throughout the world.

The periods during which the rocks of the various systems were formed are grouped into four major eras, named according to the stage they represent in the development of life. The oldest rocks, mostly devoid of fossils, were formed in the Proterozoic or Precambrian, while the oldest rocks with abundant fossils belong to the Palaeozoic Era. The geological middle ages are referred to as the Mesozoic and the era of recent life is the Cainozoic. All the systems are also known [see table on p. 91] by names which refer to places where the systems were first recognized and sometimes to special features of the rocks. The term Cambrian, for example, stems from the Roman name for Wales, Ordovician and Silurian from Celtic tribes of North Wales and the Welsh Borderland, whereas Cretaceous refers to the prominence of chalk in the sequence and Carboniferous to the presence of coal.

The beginning of each period is usually marked by a strong break in the succession of strata, breaks termed unconformities, that stem largely from the lowering of sea level relative to the contemporary land. The relative uplift of the new rocks is often accompanied by the warping or buckling of the layers—folding that repeated often enough or with sufficient intensity may produce long-lasting mountains. The pattern may be further confused by lateral or vertical movements of one part of the earth's crust rel-

5 The petrified forest, Arizona – a fossil wood in the making. The pebbly gravel around the trunks, when cemented, would make a typical conglomerate.

6 The San Andreas Fault, California – a finely detailed erosion pattern and a deep trough mark the line of this still active rent in the crust.

8 (*opposite bottom*) A volcanic plug used by pelicans as a breeding ground. Wase Rock, 100 miles south-east of Jos, Nigeria.

ative to another along a fracture or fault. The great valley occupied by Loch Ness is etched along a major fault which has allowed northern Scotland to move some 65 miles to the southwest relative to southern Scotland. Earthquakes associated with movements along the San Andreas fault threaten San Francisco. Older rocks naturally tend to be more strongly folded and disrupted than younger, and the greatest disturbance is found in the rocks which form the Precambrian basement to the fossiliferous layers.

The time required for the slow processes of rock decay and rebirth is evidently immense. Accumulations of sediment in historic times suggest that about one foot of sediment is formed in one thousand years. Geological time clearly has to be measured in millions of years. Many nineteenth-century geologists were therefore somewhat discomfited by the authoritative attempt made by Lord Kelvin to prove 'mathematically' that the age of the earth could not be greater than 100 million years. The discovery of radioactivity gave the lie to Kelvin's cogent arguments and later provided a method by which the age of rocks could be measured. It became possible to clad the geological systems with numerical data. We know now that the oldest notably fossiliferous rocks began to form about 600 million years ago, and the greatest age yet found for any rock is around 3,500 million years.

Not all earth processes are slow, however. Volcanic islands such as Sertsey off the coast of Iceland may appear overnight and at Paricutin, 200 miles west of Mexico City, a crack in the ground and a puff of ash heralded the birth of a volcanic mountain that grew by over 1,300 feet in height in a few months. Similarly constructed are the Hawaiian Islands, where molten rock originating some 50 kilometres below sea level has been repeatedly, and often catastrophically, erupted to flow on the surface. Such violently erupted material, yielding as it does the so-called igneous rocks, may appear in areas where layered rocks are accumulating. It may penetrate the strata or form sheets parallel to the layering, adding to the total bulk of new rock. Frequently the igneous masses so formed are hundreds of feet thick and extend for many miles. Because of their high temperature—often over 1000°C— they bake the surrounding materials, causing new minerals to grow in the sedimentary layers, and at the same time metamorphosing the sediments. The process of metamorphism occurs on a much grander scale deep in the crust, and whole regions of metamorphic rocks have been brought to the surface by folding and faulting. The wholesale changes in rocks brought about by heating and pressure in the depths of the crust are frequently accompanied by the growth of immense bodies of igneous looking material—the

7 Typical Lewisian gneiss. Stoer, Scotland.

granites. Indeed, it is a widely held view that many granites represent the extreme alteration of sediments and volcanic matter —a view that appears to be particularly appropriate to Precambrian rock provinces. Assemblages of granite and metamorphic rocks are often found as cores to deeply dissected old mountain chains such as the Scottish Highlands. The western mountains of the United States and South America, on the other hand, are made of sedimentary and volcanic rocks that form the envelopes to huge masses of granite without abundant metamorphic rocks.

The three broad categories of rock are as follows: the sediments, made on the surface from rock debris; the igneous rocks derived from deep within or even below the earth's crust; and the metamorphics, manufactured by heat and pressure applied to pre-existing rocks. All three groups include rocks worth collecting for their brilliance, colour or durability or structure, that provide us with a wealth of raw material both for commerce and for imaginative use.

Rocks from the depths

9 (*left*) The basalt of the Giant's Causeway, Co. Antrim, Ireland.

10 A tuff made of small crystals and rock fragments from Dolgelley, Wales.

11 An agglomerate made of large rock fragments blasted from slate near a Lower Palaeozoic volcano. Nevin, Wales.

FEW WILL be unaware of volcanoes, of the headline-raising devastation they may produce, of the searing, scorching and poisoning of vegetation that occurs as incandescent liquid rock or lava cascades over farm or village. The extinction of Pompeii is but one of the many hideous events produced by volcanicity, and the cathedral towers of San Juan Parangaricutiro standing alone above the new volcanic rocks of Paricutin appear almost as the tombstone of this town and a monument to others destroyed by the earth. But where does this terror come from? Some may, mistakenly, believe that the source lies in a molten core in the earth and many geologists only a few years ago would have pointed to the lower part of the crust as the most likely place. We know now, however, that the molten rock is first produced from not far beneath the base of the crust.

Tens of kilometres below ground at the base of the crust there is a marked change in the composition of the earth. We do not know the precise make-up of these deep rocks, but indirect evidence tells us that they are made of dark, dense material probably similar to the densest igneous rocks found on the surface. We must await the results of renewed deep drilling following up the now suspended 'Mohole' project (the drilling of a hole through the earth's crust to find out more about its geological history).

It is from within this zone of the earth, however, that most volcanic rocks are ultimately derived. Periodically large volumes the dark material partially melt, and gradually the liquid produced makes its way upwards. It may spread laterally along the bedding planes of sediments, making sills, or it may form wall-like bodies called dykes that transgress the layered rocks. Eventually it may reach the surface. Freed from the restraining influence of the enclosing rocks, it may boil off its gases, blast the solid rock into minute fragments, sending dust high into the stratosphere, or flow over the surface, devastating enormous areas of country. Dust and ashes, bombs and rock fragments are sent cascading into the air to fall in layers around the orifice. The coarsest bits land close to the volcano, often re-entering the vent itself. Finer fragments fall into beds, building the well-known conical shape, and the finest dust may be carried around the earth by the high winds miles above the earth's surface. The bedded accumulations of medium to fine pieces are called tuffs and the coarse deposits, with ovoid lumps of congealed liquid, the volcanic bombs, are termed agglomerates [figures 10 and 11]. These loose fragmental rocks

13 Obsidian, a dark volcanic glass with white clusters of microlites. Utah.

14 Ignimbrite in formation. Lava from Etna on volcanic ash, 1971 eruption.

are reinforced by dykes and sills and by occasional lava flows on and beneath the surface of the cone.

Volcanoes of this kind are built around circular perforations of the crust—the volcanic neck—and during their history frequently show signs of collapse into the mass of liquid held just below the surface. After particularly violent eruptions the whole cone may drop thousands of feet, descending about a circular fracture into the reservoir. Crater Lake in Oregon is a magnificent example of this structure. The lake occupies a steep-walled circular depression some six miles across and is sited where, about 8,000 years ago, there stood a conical mountain some 12,000 feet high. Beyond the edge of the lake are the rocks that were blown out to make room for the foundered cone. These are mostly pumice and a peculiar rock full of glassy shards called ignimbrite [figure 14]. Within the lake is a small cone called Wizard Rock that protrudes just above the water. This was built much more recently. The fine-grained loose tuffs are often difficult to cultivate, may be very prone to land-slipping, and may even flow as slurry in areas of considerable rainfall. Rocks such as these may be swept away in a sea of mud during heavy storms.

Collecting material of this consistency is of course extremely difficult, but with increasing age the tuffs become cemented and more tractable to the hammer. There are numerous occurrences

15 A typical banded rhyholite. Llanberis, N. Wales.

of this type of volcanic rock, which erupted in Ordovician times, and samples of diverse indurated tuffs, agglomerates and ignimbrites can readily be obtained by the collector. Mostly these bitty rocks are made of light-coloured minerals, though it is difficult to identify the individual species visually. The occasional solidified flows of lava are as a rule so fine-grained that the grains cannot be distinguished with the naked eye. Indeed, in many instances they are natural glass or obsidian which, when developed to near perfection, is jet black and breaks to give the typical brilliant lustre of glass. Examples are found in Shasta County, California, and green, black or transparent varieties occur as large patches in the pumice of the Isle of Lipari [figure 13]. The glass may, however, be crowded with tuff-like bits and pieces or it may be marred by minute crystallites and colour-banding that records the swirling, tumbling flow of the lava. Magnificent flow-banding is found in the lavas of the English Channel Islands and in the flat-lying flows which form the walls of the picturesque Glen Coe of Argyllshire. The flow structure may be marked by chains of small white spherical bodies made of minute needle-shaped crystals that radiate from the centres of the spheres and sometimes, in the English Channel Islands for example, extraordinary rocks occur in which the spherulites are several inches in diameter. Some inclusions in various ancient volcanic glasses are feather-like. In the United States the so-called 'snow-flake' obsidian is a good example of this.

The lavas found in company with the tuffs and agglomerates usually fall into two broad categories—the rhyolites and the somewhat darker coloured and more dense andesites. In old dissected volcanic areas it is common to find such volcanics associated with the much coarser grained granites and granodiorites which are described on later pages. The overwhelming majority of lavas, however, are neither rhyolite nor andesite, but darker rocks, more obviously crystalline with tiny light coloured laths or needles set in an almost indeterminate aggregate of dark grains. These are the basalts. They are the material of which the Hawaiian Islands are made. They build the Columbia river plateau and almost the whole of Iceland, the Deccan plateau in India and innumerable other major plateaux and oceanic islands, totalling in all countless millions of cubic miles of eruptive rock.

In contrast with the glassy rhyolites which tend to be extremely viscous and do not flow rapidly or far, basaltic lava is extremely free-moving, and lava flows may accelerate to 50 miles an hour or so on down gradients. Flows may also develop most remarkable structures as they cool—petrified structures much like those seen in running water. Gases dissolved in the lava tend to bubble out upon eruption, and many may fail to escape through the quickly cooling skin of the flow and remain as cavities in the solid rock. In extreme examples rock froth may be produced. These bubbles—technically vesicles—are commonly very slowly filled by crystals, and these filled vesicles or amygdales tend to add much to the visual attraction of otherwise fairly dull rocks.

Usually the crystals within the amygdales are the transparent, white or pale tinted zeolites. In Iceland and in Northern Ireland we find masses of really exquisite white and pale green crystals of all conceivable shapes and arrangements. There are, for example, brush-like aggregates, bladed crystals and delicate radiating needles. Apophyllite, scolecite, heulandite, stilbite, natrolite and chabazite are the names of some of the most common zeolites. Occasionally there may also be crystals of calcite, and the brilliant colour-banding of agate can be found in amygdales near the surface of some lavas. Most amygdaloidal basalts contain a certain amount of zeolite, but the volcanic areas of Northern Ireland, Iceland, Nova Scotia, Madras and Brazil are among the most excellent collecting localities. Sometimes, the vesicles of basalt, instead of filling up with these often fragile new minerals, are bounded by crystals of the rock itself, free to grow to perfection into the gas-filled cavities.

Perhaps the best known of basaltic lavas are the legendary columnar rocks of the Giant's Causeway in Northern Ireland, of Fingal's Cave on the Scottish island of Staffa and of Devil's Rockpile National Monument on the Sierra Nevada. These great tourist attractions are especially good examples of a phenomenon that occurs in basalts the world over. The roughly hexagonal columns result from the contraction of the lava flow during the slow, uniform cooling of the flow from below. Towards the tops of the lavas the columns are usually smaller, thinner and irregularly curving or radiating. Another structure found in basaltic lavas is produced where the lava flows into water, for here the molten rock breaks up into globules that roll together and finally come to rest as a pile of pillow-shaped masses. Each pillow usually has a very fine grained or glassy skin, and abundant vesicles may appear just inside this skin. The pillows are often linked together by narrow necks as though each has puffed out from its neighbour. These pillow lavas are sometimes much altered inside the globules and may contain long white needles grown randomly across all other minerals. Rocks like this are termed spilites. Some of the most extensive spilitic lavas occur in the Permian and Triassic volcanics that stretch from Alaska to south-east California—a volcanic field that has been described as the greatest of all geological time.

In areas characterized by abundant basalt, there are usually to

16 A volcanic glass with numerous very small needle-like crystals – sometimes termed pitchstone. Arran, Scotland.

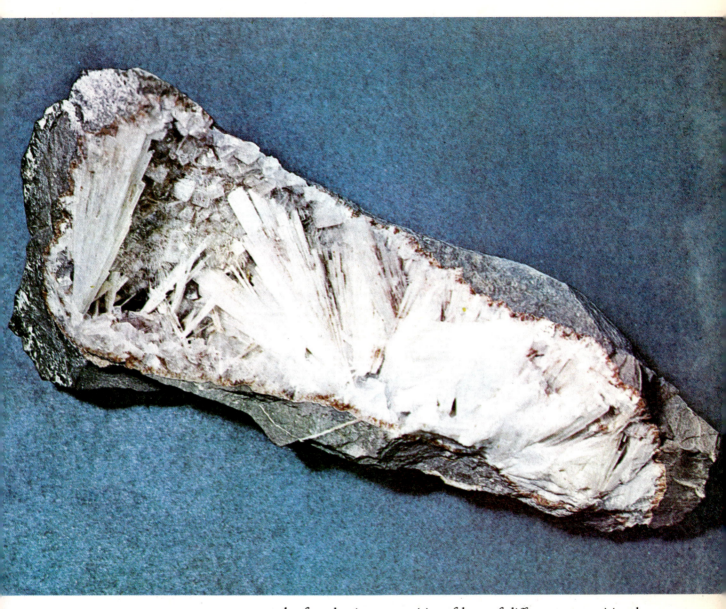

be found minor quantities of lava of different composition but thought to be derived from the basalts. The exceptionally tough and easily seen Lion's Haunch basalt of the hill of that name in Edinburgh is associated with mugearite and phonolites. The former, named after a small island near the Isle of Skye, is an example of the common geological practice of naming from small, almost insignificant, areas—a practice now deplored by most and certainly prone to cause confusion and the proliferation of unnecessary names. Mugearite is but little different from basalt in hand specimen. On the other hand, phonolite, as its name suggests, is rather harder than basalt and rings when struck, amply justifying its pet

17

name of 'clinkstone'. It is also more feldspar-rich than basalt, and is hence pale grey in colour. Trachytes are even more feldspathic than phonolite, often containing large well-shaped feldspar crystals arranged in trains that mark out the flow of the lava [figure 25]. Good examples of these rocks are to be had from Puy de Dôme in Auvergne, France; figure 24 shows a piece of trachytic phonolite of carboniferous age. Similar trachytic rocks occur in the Drachenfels area of West Germany. They are also common in the oceanic islands, Hawaii among them, and rocks intermediate between trachyte and basalt and bearing the large well-shaped feldspar crystals are found in the Highwood Province of Central Montana.

Basalts containing large white crystals of the mineral leucite are to be found in the volcanic craters near Rome, and one variety at Tavolata known as leucite tephrite is a collector's piece [figure 23]. The Leucite Hills of Wyoming are another good source locality, and similar lavas abound in Yellowstone Park among a great diversity of other volcanic rocks. Glassy basaltic rocks containing this mineral are also found at Vesuvius and in the Eifel district. Phonolites may contain other unusual minerals—such as nosean—and particularly handsome rocks are the nosean-phonolites of Burg Olbruck.

The construction of enormous basaltic lava fields obviously demands a large number of feeders to bring the fluid to the surface. Many such feeders are not circular volcanic necks but fissures extending for many miles. They form part of the sill and dyke systems that lie beneath the surface flows. The rock types of the sills and dykes may sometimes be virtually indistinguishable from the basalt flows. More often, however, they are slightly coarser grained than the lavas and are therefore termed dolerite, though they bear essentially the same minerals as the basalt. Typically in dolerites the white feldspar laths lie partly or wholly within large dark plates of pyroxene.

Some of the most spectacular dykes and sills are thousands of feet thick and may show considerable variation in composition. The giant Karroo sills of South Africa and similar bodies in Tasmania, Antarctica and Spitzbergen are examples of these varied dolerites, and one of the best known of such bodies is the Palisades sill of New York. This shows that some of the first crystals to grow in the cooling rock liquid sank towards the bottom and accumulated to give a dark 'ledge' some feet from the base. In Britain sills occur in abundance, though usually smaller than those already mentioned; but the great Whin sill of the Pennines is an exception—here is a sheet of dolerite tens of feet thick, and the sill can be traced for tens of miles. In addition, stretching for miles from the western coast of Scotland into Yorkshire are narrow doleritic dykes formed during the Tertiary period in association with the volcanics of Skye and Mull. Some of the Scottish sills are like the Palisades sheet mentioned above, having accumulations of dense minerals towards their bases.

Some of the most photogenic of all rocks are the rare composite dykes found in Skye, Mull and Northern Ireland. These usually consist of a dark doleritic margin and light rhyolitic centre. In some cases, the dolerite and lighter coloured rocks appear to have been liquid together, and most colourful and complex relationships appear between the two rock types.

Sometimes rocks of basaltic composition may be in extraordi-

23 Leucite tephrite – a basaltic lava with large white leucite crystals. Rome.

nary large funnel-shaped masses many miles in diameter. The rocks are coarser grained than the dolerite and constitute one of the most thoroughly investigated of rock suites—the gabbros. They often show thin layering with seams of dark minerals alternating with seams of light. At Stillwater in Montana, 17,000 feet of such banded rocks occur, the whole being generally darker towards the base and lighter towards the top. There are bands of chromium ore in the lower 3,000 feet, and strongly feldspar-rich rock occurs towards the top. A similar mass occurs in Minnesota and Wisconsin—the Duluth Gabbro—and one of the most thoroughly worked out masses, the Skaergaard complex, occurs in Greenland. The layered gabbros of Britain are small by comparison but this gives the advantage that much more variety can be seen in a short space of time. The gabbros and related layered rocks such as troctolites contain patches and veins of extremely coarse rock with enormous feldspars and great lustrous dark mica flakes and numerous other more exotic minerals.

Not all gabbros are layered but all contain features of considerable interest. Though they crystallized within the crust, the gabbros are to be regarded as volcanic in the sense that they are made of the stuff of eruptive rock; but the same cannot always be said of another rock suite often found associated with them—the serpentinites [figure 30]. These soft and colourful rocks are often formed by the alteration of other dark gabbroic materials. They may be white, green, red or black, or streaked and patched in all of these colours. Because of their softness and because they can be highly polished, these rocks are frequently worked into elegant ornaments or decorative objects such as candlesticks and ashtrays.

24 Trachytic phonolite, a volcanic rock from Haddington, Scotland.

25 Sanidine trachyte – the light spots are large sanidine crystals.

Bodies of serpentinite occur throughout the world and some provide a source of platinum. In the Appalachian Mountains great chains of pods of serpentinite are found, contrasting strongly in size with the small patches of the British Isles.

The serpentinites are sometimes derived by the alteration of

29 Gabbro from Cumberland, England – a coarse-grained igneous rock of basaltic composition.

30 Serpentinite – a soft colourful rock marked with the patterns that the name describes. The Lizard, Cornwall, England.

very dark coarse-grained rocks that bear much olivine. The rock takes its name from the name given to olivines of gem quality —the peridot. Some peridotites are the repository of diamond [figure 31], and since diamond requires great heat and pressure for its formation it seems likely that the rocks stem from deep in the crust. Indeed, a composition very close to that of olivine rock has been suggested as the one most likely for the rocks just below the base of the earth's crust. The diamonds may have originated miles below the present land surface.

Also rather darker than the gabbros and often found in sills and dykes are the picrites. These coarse rocks include both augite-picrite and hornblende-picrite. Gateposts and the supporting pieces of older buildings may be made from it. This tough-looking rock is almost black in colour and has a coarse-grained texture. Occasionally in a field you come across a large boulder of picrite, and a few ponds, which once were little quarries, appear to be the possible source of this rock, though no big exposures exist. Under the glacial 'drift' and peat there are obviously some good large dykes or sills, but they have not been seen or recorded since geological survey work commenced.

Volcanic areas are often well known as health spas because of the presence of natural, strongly saline, hot waters. Hot springs, boiling mud and steam jets are common, and very large areas of country may be covered with rocks that are the products of such hot springs. In western Sicily, for example, occasional exposures of rock can only be described as large selenite (crystallized gypsum) crystals in a smaller matrix. The whole is a grey-speckled white rock. Occasionally veins of sulphur and in some places beautiful salt crystals are to be found. These rocks are something quite out of the ordinary. In some places the rock is calcareous tuffa or travertine [figure 32]. Much of Rome is built of these rocks. You can often see where snail shells and vegetation have been caught up during the formation of the rock.

There exist side by side in a certain part of the English Lake District two very large quarries. One hole is blasted in andesite lavas erupted in the Silurian period. These are dark fine-grained purplish rocks streaked with white and with thin seams of minute but perfectly formed clear green crystals of the mineral epidote. The rocks have been somewhat metamorphosed, and the cause of this baking is immediately revealed on a visit to the second quarry. Here one of the world's most famous ornamental granites is being excavated. This granite is characterized by large rectangular pink feldspar crystals that are often set in lines that twist and turn as though the crystals were carried in a thick flowing liquid. Filling the space between these crystals is a coarse mosaic

31 Peridotite, a dense diamond-bearing rock made of dark minerals – principally olivine. Kimberley, S. Africa.

of glassy quartz, black lustrous plates of mica and smaller pink and white feldspars. The rock shows very little variation throughout the quarry, but here and there are found black spots made of very fine-grained rock much like the nearby andesites. These spots are most simply regarded as xenoliths, as bits of andesite stripped from their original site by the liquid granite as it made its way upwards through the crust.

One of the most intriguing aspects of the xenoliths is that they frequently contain large pink feldspars identical with those in the granite. Sometimes so much feldspar has developed that the black mass is reduced to a vague patch—a ghost of its former self. The granite can alter the inclusions and itself become contaminated with the material of the dark masses.

In some parts of the granite there are sheets up to several feet thick of uniformly fine-grained pink granitic rock termed aplite, and these are often associated with sheets and blebs of extraordinary coarse granite known as pegmatite. The crystals of pegmatites are frequently measurable in inches, and such rocks often contain concentrations of many of the minor constituents of normal granite. These rare constituents make some of the most spectacular and sought-after crystals such as garnet, tourmaline and beryl, zircon and topaz.

The association—granite-aplite-pegmatite—is found throughout the world. The granites differ [figures 35 and 37] in texture and the pegmatites in the size of their crystals and in the varieties of rare minerals present. The quantity of xenoliths and the degree of contamination of the granite also varies from place to place. Large feldspars sometimes occur, and sometimes tourmaline.

32 Calcareous tuffa, made of calcite grown in concentric layers.

Radiating clusters of the lustrous black mineral may be set in a background of pink feldspar. The granites may in places be much decomposed by gases emanating from the depths, the feldspars rotted and converted to fine white china clay.

Granites are often surrounded by hosts of small dykes, and fine-grained dyke rocks with large quartz or feldspar crystals can often be formed in tin-mine areas [figure 38]. Dykes called lamprophyres often contain extremely well-shaped and brilliant mica plates. But rarely are they undecomposed collectable rocks. Another very striking solid rock is the hornblende-lamprophyre called vogesite, which displays green laths of hornblende in a pale ground-mass.

The contacts of the various granites with the sediments are often steeply inclined and it appears that the bodies become larger downwards. Indeed it has been found that all these masses are interconnected not far beneath the present land surface. Geologists have for generations debated as to how such volumes of rock can have been introduced into the upper levels of the crust and where the original sedimentary rocks have gone. One explanation came from examining the granites of New Hampshire. Here the margins of the igneous rocks are replete with large xenoliths of the surrounding rocks, and many enormous blocks look as though they were just about to be stripped from the walls of the

34 Typical granite. Dartmoor, Devon.

35 A granite made of large rectangular white feldspars, grey quartz, dark mica and tourmaline crystals. Dartmoor, England.

36 Axinite rock, in which large crystals of the mineral axinite have grown as a result of the introduction of boron from nearby granite.

37 Tourmalinized granite, Cornwall, England, of a type similar to that in figure 35, but altered by the growth of large amounts of tourmaline. The large white feldspars are not altered and hence stand out clearly.

granite. The wholesale sinking of such blocks could permit the passive uprise of the fluid granite. More recent work on these rocks has shown that the granites form near-vertical rings, and it seems likely that huge cylinders of sediments have subsided into the granite cauldron.

Large intrusions of granite are termed batholiths; the greatest of them all stretches from the Andes through California into Western Canada. The mountain country in which the batholiths occur is usually hostile, and much remains to be done in mapping this complex. Already, however, it is clear that the batholith is made of hundreds of relatively small intrusions all differing in composition and texture. In Peru and southern California, for example, much of the rock is not true granite but granodiorite, with white calcium-bearing feldspars and more dark mica and less quartz than the true granites. There are also spectacular rocks made of large, dark, lustrous prisms of the mineral hornblende and white feldspar—the diorites. Coarse gabbros made of pyroxene and feldspar grade into the diorites, and more rarely there are very dark rocks entirely devoid of the white feldspar. The light-coloured granite rocks often form veins in the much darker diorites and gabbros and dark blocks abound in the lighter rocks —forming some of the most photogenic and picturesque of igneous rocks.

Many small batholiths are found in Scotland and Ireland and show structures and rock varieties very similar to those of southern California. In Scotland there is an immense variety of rocks, including spectacular, lustrous, black crystalline masses shot through with white veins and masses of sediment blasted into fragments by the intrusion of the igneous rocks.

A striking rock that has an unusual assortment of minerals is the kentallenite from Kentallen in Argyllshire, Scotland. This is a relative of the diorites that so commonly accompany the granites of Scotland and elsewhere. Diorites may occur as small bodies in their own right, as marginal rocks to the granites, or sometimes as rafts in the granites or granodiorites. Similarly a rock much resembling diorite, though characterized by pink feldspars, is also found associated with granite. This is termed syenite. At Plauen in East Germany there are complexes containing both syenite and granite. A highly iridescent syenite obtained from Sweden is often used in polished sheets for the exterior panels of shops.

Though syenites are found commonly with granites, they also occur prominently in another association—with a remarkable group of igneous rocks called carbonatites. Syenitic rocks in this assemblage are often handsome and very diverse in their textures

38 A granitic dyke rock with medium-sized crystals in which are set large quartz and feldspar crystals. Cornwall.

and mineral proportions, and many special names such as ijolite and foyaite are used to describe them.

The carbonatites, as the name implies, are carbonate-rich rocks of igneous origin, and some years ago these were regarded as limestone caught up in the other igneous rocks. They are not limestones, and it is probable that they were originally formed from igneous liquid containing much carbon dioxide gas. The liquid intruded into other rocks that surrounded it, forming dykes and sills. More recently volcanic lavas and coarse tuffs of carbonatite rock have been found.

A great range of interesting accessory minerals is often found with the carbonatites, and many have yielded minerals of economic importance. The rare element niobium has been found in some areas, and carbonatites have been discovered on the Isle of Alnö in Sweden, in the Monchique area of Portugal, at Magnet Cove in Arkansas, at Fen in Norway, and in countless other parts of the world, particularly Rhodesia, Tanzania and Malawi. Indeed it was some of the African examples that first made apparent the importance of this remarkable suite of rocks.

39 (below) Orbicular diorite, an unusual rock in which the white feldspar and dark hornblende of the diorite form distinct concentric spheres. Finland.

40 (right) Borolonite from N.W. Scotland – a syenite with large white spots of feldspar, nepheline, dark mica and garnet.

41 (below right) Orthoclase porphyry – a syenitic dyke rock with large pink feldspars in a feldspar-rich background of medium-sized crystals. Scotland.

31

Rocks from salty seas

42 (*left*) A mountain of salt at Kuh-i-Namak. This is a salt-plug forced upwards through overlying rocks; buried structures of this kind commonly make oil traps.

43 A salt plug at Murfadil, Iran. The bedded sedimentary rocks are arched up by the forceful intrusion of salt.

BETWEEN Abyssinia and Eritrea where the Red Sea begins to narrow, there are on the land surface thick masses of salt. The Red Sea once covered this area, but its access to the region was cut off by eruptions from a series of volcanoes. The territory is one of the hottest and driest places on earth, and not surprisingly the pool isolated by the volcanoes rapidly evaporated leaving behind thick saucer-shaped beds of salt.

Similar beds of salt have been formed from time to time throughout the history of the world, and are especially abundant in Permian and Triassic rocks. The climate of the whole world must have been much hotter and drier then than it is now, for the salt deposits are extraordinarily widespread. These deposits are of considerable industrial importance for, apart from our dietary needs, they provide some of the most prolific source rocks for oil. The salt beds are much lighter than other rocks and therefore tend to be squeezed up into cracks in overlying beds. Once the salt begins to rise it is squeezed by the overlying load so that it continues to rise and builds great salt columns. It is these columns, or 'salt domes', that make the most important oil reservoirs. Salt domes abound in the United States, especially in Texas and the Gulf Coast states, and in Germany and other parts of Europe, and it is salt beds that provide the tanks for the great North Sea gas field.

Many of the deposits are made of sodium chloride—the common cooking salt. These beds are referred to as rock salt and the mineral that makes them is called halite [figure 44]. Mines in Poland, Germany and Alsace-Lorraine have been worked for hundreds of years. The halite is often found in white, pink or clear cubes that may be several inches across. In New York State there are thick deposits of rock salt, and in central Michigan layers of rock salt are accompanied by gypsum. Clusters of water-clear crystals are found in salt deposits in Sicily and Poland, and in the famous deposits at Stassfurt in Germany the crystals are sometimes a delicate blue.

At Stassfurt, there are also extensive beds of carnallite and sylvine which are potassium salts used to make fertilizers. Carnallite and associated minerals are also worked in the Dead Sea area.

Salts are among the softest of minerals, yet in many instances they make excellent water-clear crystals that are well worth preserving. Gypsum, for example, forms good crystals. This mineral has water in its structure, and if powdered and heated the water may be driven off and the gypsum converted to plaster of paris. The Triassic rocks of England and Wales frequently contain gypsum, and Oklahoma is well known for its gypsum beds. The mineral often forms irregular masses in clay and may take on a delightful pink colour when slightly iron-stained. In some places it forms long slender parallel fibres. It has been much prized and used for monuments, tombstones and bijouterie for hundreds of years, and in Italy there are splendid specimens of a semi-transparent form of gypsum called alabaster. In Pisa it is common enough to see men carving the alabaster into small replicas of the leaning tower for sale to visitors; this variety of gypsum can be coloured artificially and carvings sold in Mexico as jade are often dyed alabaster.

Gypsum is often accompanied by the very dense mineral celestite [figure 45] which forms fine crystal masses, and pockets can be found lined with the white and translucent crystals. The basic substance of gypsum, but lacking water in its structure, occurs in many salt deposits and is termed anhydrite; layers of anhydrite abound in Texas and New Mexico. It is a yellowishwhite banded mineral that often occurs in kidney-shaped masses.

A note on making microscope slides of rocks

Normally rocks are opaque and must be viewed under the microscope with reflected light. But with care rock chips can be ground so thin that they become transparent and can be viewed with transmitted light. Standard kits can be obtained for making rock slides but all you really need are a few thick glass plates about a foot square and three grades of grinding paste. The glass plates

44 Water-clear rock salt crystals. Sicily.

should preferably be slightly convex. Each one needs a small wooden tray to keep it clean and separate from the rest.

Apply a thick paste of coarse carborundum to one glass plate and grind a flat on the rock chip about an inch square. Wash well and smooth the flat face with medium abrasive on the next glass plate. Again wash well and polish the flat chip with the fine finishing abrasive on the third glass plate. Stick the chip, flat side down, on a square of glass with an adhesive medium such as Canada balsam. Grind the exposed rough face with coarse, medium and fine abrasive in turn until the section of rock is one thousandth of an inch thick, so thin that quartz appears yellowish-white and the feldspars grey when the slide is viewed between polarizing plates. Warm the glass square to melt the Canada balsam, and carefully float the thin film of rock on to a 3 in × 1 in microscope slip. Coat the rock slice with warm Canada balsam and put on a cover glass. When cool, clean the slide with methylated spirits and label. It is then ready for study under the so-called petrological microscope.

The petrological microscope incorporates two pieces of polaroid—one between the eyepiece and the objective and one beneath the microscope stage. These are carefully set so that when there is nothing on the stage the polaroid will cut out all light and the field of view will be completely dark. If the thinly ground rock slice is inserted when the polaroid pieces are in the dark position, the minerals on the slide will show very attractive natural colours and will be immediately identifiable. Coarse, fresh, undecomposed rocks make by far the best thin rock sections.

45 Soft, dense, clear crystals of celestite, a strontium mineral. Bristol, England.

Rocks that are not what they were

JAMES Hutton's dictum that the present is the key to the past breaks down when we have to interpret rocks that are made deep in the crust. We have no chance of witnessing the growth of these rocks, and can but imagine the processes that have gone on. What we do know is that deep mines get warmer the deeper we go—that the temperature of the earth increases as we go down by between 10 and 30° centigrade per kilometre. Again, little imagination is required to realize that, as rock is piled upon rock, an ever-increasing pressure will be applied to the deeper layers. We can also be sure that, if rocks have been folded, they must have been squeezed or stressed in some way. All these factors promote changes in the minerals and in the structures of the igneous and sedimentary bodies.

Limestone is burned in kilns to make garden lime, and in this process carbon dioxide gas is driven from it. If, however, a limestone is heated deep in the earth, the pressure of the overlying rocks prevents the gas escaping. Just a little carbon dioxide is driven into any pores or spaces that may be present, and this small amount of gas helps to start far-reaching changes in the rock. The tiny chalk grains or the shelly fossils begin to be changed into clean, clear, uniformly-sized crystals of calcite. Eventually, given enough heat, the limestone will lose all trace of its original variations and become a homogeneous mass of calcite crystals—a marble. An exceptionally pure limestone will be converted into an unadulterated gleaming white marble such as the famous monumental stone of Carrara in Italy. Similarly, a sandstone may when deeply buried become recrystallized into new quartz grains all interlocking with one another to make the exceptionally hard and resilient rock quartzite.

Changes such as this can convert a dull, featureless sediment into a magnificent coarse crystalline aggregate. Usually, however, the first mild changes, induced by gentle heating and the squeezing that makes folds and faults, produce no spectacular change. Tiny flakes of mica may grow at right angles to the stresses that cause the folds, and if this happens on a large scale the rocks become easy to split along the layers of platy minerals. They are converted into slates. The great quarries of the rugged hills around Snowdon in North Wales have provided the roofs of millions of houses. The rocks altered were very fine-grained shales and volcanics, most of Ordovician age, and it is the Ordovician black shales that have provided most of the slate in America. Here the

37

47 Phyllite, a metamorphic rock made predominantly of green 'micas' in the soft layers and quartz in the harder bands. Loch Leven, Scotland.

48 Schist, a metamorphic rock made mostly of mica, quartz and feldspar in bands. The crinkling of the bands testifies to the stresses that must have been applied to the rock during and after the growth of the metamorphic minerals. Scotland.

more homogeneous beds are split into shingles for roofing or electrical switchboard slabs, the chief producing areas being Pennsylvania, Vermont, New York and Virginia.

If the squeezing that produces folds is very intense, the rocks may split and one block may be carried over another for many miles along the plane of fracture. Rocks close to the fracture may be smashed or ground into fine dust, and the friction may be so great that enough heat is produced locally to melt the various rock types, producing the banded ultra-fine-grained or glassy substance known as mylonite.

Rocks buried deeper in the crust and subjected to higher temperatures change much more dramatically. The micas are no longer small but grow to large flakes all arranged in interweaving bands. When the grade of alteration is relatively low, the platy minerals tend to be green and the rock is termed a phyllite [figure 47], but at higher temperatures the thorough-going changes produce brown or white micas in interweaving layers alternating with light coloured seams of quartz grains. These rocks are the schists, which form the basic types of large areas of metamorphic country. Remarkably small, very sharp folds are frequently found in these rocks, and the whole contorted, crumpled mass testifies to the power of the forces affecting the metamorphism. Crinkled schists to be found in the western part of this Highland belt exemplify some of these structures [figure 48]. The lower slopes of the Grand Canyon—towards its western end—display many spectacular crystalline schists, and magnificent samples can be obtained from Monroe, Connecticut. It is often extremely difficult, when changes of such an extent have occurred, to recognize the original sedimentary bedding of the rocks, and other structures, both sedimentary and igneous, may be blotted out. Metamorphic rocks themselves may be involved in a second phase of change, and in some cases there is evidence of several separate metamorphisms.

Special minerals appear in schists, each tending to reflect the conditions of heating or pressure to which the rocks have been subjected. The Connecticut schists already mentioned contain some large, perfectly shaped garnet crystals along with the micas and quartz, and garnets abound in the schists around Pitlochry and Garth in the Scottish Highlands. More rarely the schists may contain long white or pink needles with square cross-sections of the mineral called andalusite. Rocks of a somewhat higher grade may contain the brilliant blue bladed crystals of kyanite. The giant near-perfect crystals of this mineral found near St Gottard, Switzerland, are only rivalled by those of Phoenix, Arizona.

Accompanying the kyanite are sometimes large golden staurolite crystals, and at the highest temperatures white needles of

sillimanite may be found. Staurolite frequently occurs in the form of a cross with the crystal grown as though one is penetrating the other. It is extraordinarily resistant to wear and decay and is often left perfectly preserved on the land when the original schists have rotted and been made into soil. Farmers in Virginia and North Carolina turn them up with the plough, and in the mountainous country of New Mexico crystals can be picked from the surface. All these special metamorphic minerals make excellent gemstones, for they are extremely hard, and often colourful and brilliantly reflecting. The staurolite crosses and some of the andalusite crystals are also used as amulets.

In areas showing the very greatest changes in rocks, granite patches often occur, and the schists grade into areas of banded rock in which dark bands rich in brown mica alternate with thin pink granite layers. These colourful rocks are the gneisses. The coarse quartz and feldspar of the granite fraction sometimes form eye-shaped blebs distorting the banding, and often the bands may be sharply crumpled or thrown into numerous swirling folds. Gneisses form the bulk of the Precambrian foundations of most of the continents. Enormous areas of north western Scotland are made of such rocks now exposed in low hummocky hills that are almost stripped bare of vegetation. In places the banded gneisses are fine-grained and granular and contain garnet crystals several inches in diameter. Elsewhere extraordinary knots of very long green crystals of hornblende occur in the gneiss. The enormous

49 Staurolite schist, a metamorphic rock in which large golden-brown staurolite crystals have grown in response to moderate temperatures and pressures. Norway.

50 Epidiorite, a gabbro that has been metamorphosed so that the original feldspar and pyroxene are changed to give a new feldspar and hornblende. Cornwall, England.

51 Spotted slate from Cumberland. A metamorphic rock in which spots of cordierite have begun to grow because of the high temperature imposed by a nearby granite.

52 Chiastolite slate, a metamorphic rock in which needle-shaped crystals of chiastolite have grown because of the high temperature of a nearby granite. Cumberland, England.

Precambrian areas of Canada have an abundance of gneiss and the foundation of the Grand Canyon includes similar rocks. The metamorphic country of New England and parts of the Rockies of Colorado and New Mexico also include gneissic rocks.

We have far from exhausted the list of new rocks made over large regions by the metamorphic processes. What of the gabbros, for example? These may be changed to hornblende-feldspar rocks that much resemble diorites and are sometimes called epidiorites [figure 50]. These rocks may also contain garnets, and at high temperatures and high pressures they may be converted into spectacular rocks called eclogites made of red garnet and green pyroxene. Other very high-pressure rocks are exemplified by schists found in the coast range of California. These contain brilliant purplish bladed crystals of the mineral glaucophane, and jadeite is found in the same association.

Another source of heat that causes considerable change in rocks stems from the uprise of the large masses of granite into the higher crustal levels. These molten rocks are a at temperature of some 700 or 800°C and cool down over hundreds of thousands of years. Rocks around such hot bodies are usually altered in zones, with the greatest changes occurring closest to the heat source. A good example of this graded sequence is found in the neighbourhood of the Skiddaw granite in the English Lake District. Here, as one approaches the granite, it becomes evident that the slates of the area are spotted with spongey-looking, almost indeterminate, markings [figure 51]. Closer to the granite these spots become flecked with brown and white and closer still they become distinct micas. It is at this stage that the full glory of the metamorphism begins to be recognized. The rocks gradually become a granular mosaic of micas, quartz and assorted special minerals. In some places long white crystals of andalusite appear, and if a sample is obtained showing the square cross-sections of the crystals it will be found that they contain minute black inclusions arranged in the form of a cross. The name chiastolite is given to andalusite when it contains this pattern [figure 52], and in other parts of the world fine specimens of chiastolite are cut to reveal the cross and mounted as lucky charms.

Such metamorphism can be found the world over wherever there are granite masses. In Idaho and New York, in the Adirondacks and in the Rockies there are granites and their associated metamorphic envelopes. In Scotland and Ireland too, zonal development of the new minerals can be found around the granites, sometimes with crystals several inches in length. But by far the most spectacular rocks are the altered limestones, especially where they are very close to the granite. Green or pink garnets and perfectly formed amber or brown idocrase crystals, all often extremely large, grow together with countless rare and handsome substances making colourful marbles [figure 53]. The famous Connemara marble of Ireland is white streaked with yellow and green. It is termed in general ophïcalcite marble, and is made of white calcite and green olivine or peridot and olivine altered to yellow serpentine minerals.

Gabbros also effect contact alteration, and again it is the limestones that develop the most colourful assemblages. Gabbros also have a profound metamorphic effect on shales and slates, and may even produce melting very close to their contacts with the surrounding rocks, so that glass is produced in the altered rock. Rocks containing large red garnets and big black cordierite spots are often developed in gabbros that have been heavily contaminated with shale or similar rock [figure 54].

Still we have mentioned but few of the great range of new rocks that can be made by the processes of change, and any metamorphic area will provide the collector with an embarrassment of riches to select from. A range of coarse crystalline rocks with most complex and elegant structures is to be often found in juxtaposition over very small areas.

53 Bloody marble, an impure limestone which has been metamorphosed by an igneous mass to yield coarse white calcite and dark red-brown hornblende. Scotland.

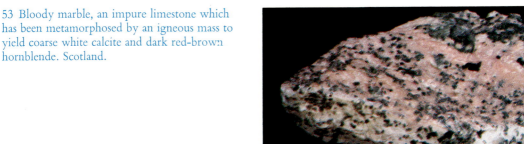

54 Norite, a contaminated gabbro containing metamorphic minerals. Huntly, Scotland.

Rocks as history books

55 (*left*) A Jurassic swordfish spears its prey – but both died, to be preserved in fine detail in the limestones of Bavaria.

THE HISTORY of rocks has no clearly defined beginning, and we cannot yet predict the conclusion. But each of the geological periods represents a chapter in the story of the earth, showing the stages in the evolution of life and the development of the land masses as we know them today. We can begin with the chapter read from the Cambrian rocks, for it is these beds that contain the first abundant signs of life.

During Cambrian times Europe, Africa and the Americas were almost certainly joined, and there was no Atlantic Ocean as we know it. The bulge of north-west Africa filled the hollow of the eastern outline of the United States and the embayment of south-west Africa was fitted against the protruberant north-western corner of South America. Western Canada, Greenland, Britain and Scandinavia formed the northern part of one vast continent. When Cambrian times opened, a long inland sea began to develop that soon stretched through Norway and Sweden, across Britain and through into what is now Newfoundland, and thence along the present eastern seaboard of the United States through the Appalachian Mountains district towards the Gulf of Mexico. A second inland sea grew along a north-south line from lower California to British Columbia and Alberta. These two great hollows were eventually linked through Ouachita, and the sea spread to cover shallowly large parts of the United States south of the Great Lakes.

If we had flown across America along, say, a line through San Francisco and Washington, DC, we would have passed first on the Pacific margin a range of mountains, then the western Cambrian seas and over a central lowland before reaching the eastern Appalachian Sea and then another mountain range. The destruction of these mountains provided material for dumping into the contemporary seas, and we can find today the coarse pebble banks and thick sands that represent the first influx of material into the water. In Middle Cambrian times the amount of fragmental material thrown down was reduced, and a warm climate suitable for the development of abundant shelled creatures and corals prevailed. The calcareous skeletons of these animals built a great thickness of limestone on the Cambrian sea floors—deposits that are now indurated and thrown up into the great mountains of Appalachia and Western America. Those willing to attempt such snow– and ice-mantled peaks as Mount Lefroy of the Canadian Rockies will find huge cliffs of these Cambrian limestones, and

56 Breccia, made of angular fragments cemented by fine sand. Cornwall.

57 Conglomerate from Hertfordshire, England. A typical 'fossil' beach in which well worn, rounded pebbles are cemented together by sand.

Mount Bosworth reveals a thickness of more than a mile of strata. The shallow seas that spread over much of the Mississippi area gave rise to sandy formations now exposed in Minnesota and Wisconsin, while the sandstones of Keezeville, New York, represent the detritus poured into the Cambrian seas of Appalachia.

Meanwhile in Britain coarse pebbly deposits, finer sands and extremely fine silts and muds were poured into the seas, yielding the conglomerates sandstones, siltstones and shales we can easily find today in Wales and Scotland. The creatures living at this time were not crude, simple forms but advanced animals such as the insect-like trilobites [figure 58], some of which were about a foot from head to tail. These flourished in the Cambrian seas and survived until about 200 million years ago. Numerous kinds of shellfish existed, such as *Lingulella* which built banks [figure 59], and another called *Lingula* has survived unchanged to the present day.

In both Europe and America the end of the Cambrian was marked by gentle earth movements that made land where the seas had been, but the emergence was short-lived and soon the seas had spread back. In Britain they occupied much the same areas as in the Cambrian, but in north America water spread over more than half the present continent, reducing it to a series of islands that were almost awash. The seas were warm and clear and widespread limestones with identical fossils throughout the country were formed in them. In New England and the coastal provinces of Canada, however, the seas gave rise to thick, dark clay deposits that eventually became black shales, and within these black shales are to be found the fossils of colonial creatures, looking more like pencil marks than animals, called graptolites [figure 60]. In both Britain and the United States, regular outbursts of volcanic rocks poured into the seas and built chains of island volcanoes. In Britain the Ordovician closed with mild folding, but in America from Newfoundland to Alabama a chain of mountains grew that stood as land for many millions of years. The Ordovician fossils are all marine, and there is no sign of land life, but there are a few traces of the bony armour of extremely primitive freshwater fish.

The sea was not long in returning to cover North America, and soon the Silurian rocks were being formed, the Ordovician Appalachians were destroyed, converted into sand and mud in the seas, and new volcanoes appeared in eastern Maine and New Brunswick, building layered sequences of some 10,000 feet in thickness. The seas became silted up and a landlocked, shallow, salty lake, stretching from the Great Lakes towards the eastern coast, remained. This gradually dried up to produce vast deposits of salt mixed with fine muds that were slowly being fed into the

58 Slate with trilobite fragments. Bohemia.

59 Moulds of the shell of *Lingulella* in slate. N. Wales.

60 Pencil-mark-like colonies of the minute graptolites typical of the Ordovician period. On slate, from Wales.

receding water from the low surrounding land. So ended the Silurian in the United States.

In Scandinavia and Britain the contrast was remarkable. The sandstones, limestones and shales of the Silurian gradually infilled seas not markedly different in position from the Ordovician and Cambrian waters. The fossils in the shale are still of graptolites, but shells and corals abound in the limestones, and the soft coral limestones of the Welsh borderland are excellent sources of fossils [figure 61]. But it is the end of Silurian times in Scandinavia and Britain that contrasts most strongly with that of America, for here the whole thickness of Cambrian, Ordovician and Silurian strata was crushed and buckled into great fold mountains. Huge belts of metamorphic rocks were made in Scotland, Norway and Sweden, and enormous granite masses made their way up through the bedded rocks. As the mountain-building movements began to wane, deep chasms and hollows were excavated and then filled with great boulders, pebbles and sand derived from higher ground. The magnificent boulder beds and conglomerates of the far northeast of Scotland, resting upon the steep surfaces of deeply weathered and etched granite and metamorphic rock, paint very clearly a picture of extraordinarily rapid deposition [figure 62]. These beds are known as the Old Red Sandstone and were formed in the Devonian period. During the accretion of the coarse beds, temporary water holes allowed finer bedded sediments to be formed, and primitive fish lived and died in the pools and became entombed in the fine sediments.

Across southern England and southern Ireland, on the other hand, stretched a broad sea in which coral limestones, sandstones and shales were produced—rocks which include the first Devonian beds recognised as such. In Britain the mountain-building shifted the positions of land and sea, but in the United States the Devonian waters began to develop along the same lines as the Cambrian seas. The area under water gradually spread over the low land mass, first extending towards the Mississippi Valley and then joining up with another deep trough on the line of the westernmost Cambrian sea, to make a seaway more than 1,000 miles across, though mostly shallows in the central parts. The rocks formed in these seas are spectacularly shown in New York and Pennsylvania. Limestones grade laterally into sandstones as the old land surface is approached, and thick sandstones were formed on top of all these rocks. Towards the top these sandstones contain traces of land plants and fresh-water fish.

During later Devonian times the Appalachian Mountain chain began to grow again, and by the end of the Devonian a large mountain system followed the axis of the old Appalachia. As in

45

Britain, the folding was accompanied by much volcanicity and the formation of granites now revealed in the core of the Devonian mountains. The life of Devonian times was remarkable. Carnivorous armour-plated fish reaching twenty feet in length dominated the diverse aquatic world, and trilobites scurried on the muddy sea floor between the long slender stalks of animals—crinoids—that looked more like plants. Distant spiral-shelled relatives of the modern pearly nautilus preyed on the trilobites, and there were countless corals and shells. But most important were the first four-legged land animals, forms that were much like fish but had very short, inefficient, five-toed legs. On emergent land grew forests of scaly primitive trees, and in seasonal ponds lived the fish ancestors of the later amphibians–fish like those surviving in Africa today, that behaved as normal fish in wet seasons, and either hibernated in mud or were able to breathe air directly when the pools became shallow and stagnant.

So far, some two hundred and fifty million years have been accounted for, and we have but another three hundred and fifty

61 A typical compound coral (heliolites) from Silurian limestones.

62 Coarse bedded deposits similar in appearance and composition to Devonian conglomerates. Wadi Bull Bull.

63 Gordale Scar, England. Typical carboniferous limestone country with thin soil and few trees.

64 A limestone full of corals. Derbyshire, England.

million years to bring the story up to the present. Again we find the sea flooding over the old Devonian land mass, coral seas building the great limestones of the Mississippi Valley area that are the most widely used building stones of the United States [figure 63]. Similar warm seas covered much of Britain, and again were the generators of immense quantities of limestone packed with fossils of colossal variety. These beds are the lower deposits of the Carboniferous, and in America they have been given the status of a system named the Mississippian.

At the end of these early Carboniferous times the sea began to fill up and the low surrounding land was periodically inundated with water. Great deltas built out into these shallow seas and at times, as the sea temporarily withdrew, great forest of primitive trees developed and enormous thick succulent vegetation grew in large swampy areas. The fine structure of some of this vegetation is marvellously preserved by petrification in the limestone knobs known as coal balls [figure 66]. Dragonfly-like creatures with wings a yard or more in width hovered over slithering giant amphibians in these hot jungle-like terrains. Dying plants and trees collapsing into the swamps built great thicknesses of only partially decayed vegetation upon which new forests continued to flourish until once more the sea inundated the areas and new masses of sand were piled upon the fetid swampy morass. This sequence of events, repeated time and again, has provided us with the great coal series of the upper Carboniferous in Europe and

Britain and the equivalent Pennsylvanian period of the United States.

The Devonian mountains of Appalachia were not destroyed by the Carboniferous seas, and indeed in Middle Carboniferous times the Appalachians were extended and new mountains grew in Colorado; all these earth movements culminated in the uprise of huge world-wide mountains during Permian times. All the continents were emergent land areas by this time, deserts were widespread, and one of the greatest spreads of ice and snow took place. For many of the Palaeozoic life forms, the fantastic changes in land-sea distribution and the abrupt climatic changes sounded the death knell. The signs of increasing aridity are unmistakable. The retreating warm seas were margined by coral reefs, and eventually dried to give huge salt deposits in, for example, the Guadalupe Basin. In South America and South Africa, on the other hand, great ice sheets spread over the land—strong evidence indeed that these continents were linked at the time and that both must have been much nearer to the polar region.

Reptiles—land animals that were for almost the first time entirely freed from the water—found the warmer climate comfortable and began to flourish. Great lizard-like creatures, with huge bony dorsal fins, hunted and scavenged through the dry, poorly vegetated desert.

These desert conditions prevailed throughout the succeeding Triassic times; the mountains built in North America during the Permian were abraded and rocks much like the Devonian red beds were formed. Similar red sandstone carpeted much of Britain, often containing pebbles that were wind-blasted into wedge-shapes called dreikanters [figure 69]. In Newark, New Jersey, and several other eastern states, however, great faults allowed huge parts of the Appalachian chain to drop several thousand feet, and deep basins were formed into which poured ton upon ton of debris to make conglomerate, sandstone and shales. In western

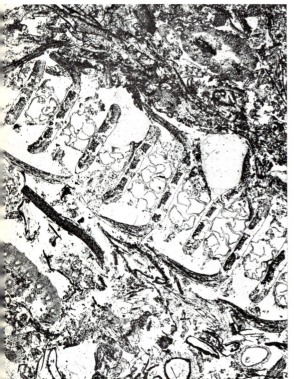

65 (*top*) Limestone made of fragmented crinoid stems and solitary corals. Derbyshire, England.

66 (*above*) A microscope section of a seed cone, obtained from a coal ball found in Lancashire, England, showing the quality of detail that is often preserved by petrification.

67 (*right*) Foliage (*Alethopteris*) perfectly preserved in shale from the coal measures of Yorkshire, England.

48

68 (*right*) Footprints of a Triassic amphibian, found in Cheshire, England.

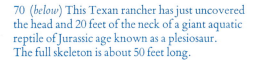

69 A wind-faceted pebble from the Triassic rocks of Derbyshire, England, testifies to the desert conditions that prevailed during that period.

70 (*below*) This Texan rancher has just uncovered the head and 20 feet of the neck of a giant aquatic reptile of Jurassic age known as a plesiosaur. The full skeleton is about 50 feet long.

71 (*below right*) Dinosaur fossils dug up in the Maragha Valley, Iran, by the American Museum of Natural History.

Europe the uppermost Triassic beds—sometimes referred to as the Rhaetic—contain fragmentary remains of the first tiny mouse-like mammals, but the Triassic rocks have more usually yielded giant reptiles like crocodiles and others that were more like mammals. Vertebrate bones occur in the Rhaetic beds at Aust Cliff in Gloucestershire, England, where they are found with fish and reptile teeth.

It is the Jurassic rocks that have given some of the most spectacular reptile remains. The Jurassic is not widespread in the United States, but a group of shales, silts and sandstones called the Morrison formation, stretching from Montana into Utah and then south to New Mexico, has yielded magnificent remains of the

73 Ammonites, distant relatives of the pearly nautilus preserved from Jurassic times. England.

great dinosaurs. These beds were formed from streams winding over a large flood plain with many small marshy lakes and pools. The brontosaurus was more than 60 feet from head to tail and nearly twenty high, and was but one of the sixty-odd species of dinosaur found in the deposits. The savage carnivorous allosaurus was about 35 feet long, and these great beasts thrived in the warm climate and often spent much or all of their lives in the shallow waters—to take the weight off their feet. Some reptiles anticipated the birds, and the remains of flying beasts with wing spans of 20 feet have been found. Other reptiles returned to the seas, and the warm shallow waters that had by this time flooded over much of Britain swarmed with the long-jawed ichthyosaurs. Remains of these animals have been found in the Jurassic strata of southern England. Another kind of animal remains consists of small concentrically shelled pellets of chalk, often built around bits of shell or sand. They are called oolites.

There were at this time many other types of backboned animals including a few mammals; it was during the late Jurassic times that the first birds entered upon the scene, but all the fossils of these forms are rare. Much more prolifically preserved are the diverse coiled shells called ammonites that are much like the modern pearly nautilus. It is these fossils that are most often used to identify the different rocks of the Jurassic and they may be collected the world over. There were also the cone-shaped belemnites [figure 74], and reefs of sponges and corals. At Solenhofen in Bavaria, such reefs have been quarried and sent all over the world for engraving and for use as lithographic stone. The Solenhofen beds also include worm-tracks, lobster-like forms and, occasionally, fish [figure 55]. But the most splendid of all are the fossils of primitive birds [figure 76]. They show many stages in the development of the bird from a reptilian, bat-like form to the archaeopterix, the first true bird. Sometimes these fossils fluoresce and show striking patterns under ultra-violet light. A marvellous display of the fossils of this important rock is housed in the Vienna Natural History Museum.

Similar animals survived into the Cretaceous period which marks the last great advance of the sea. In these waters was formed the famous chalk of southern England and from seas of this age arose the majestic Rocky Mountains—the whole chain from Alaska to the Andes. Within these seas lived delicate spiny sea urchins, much like those found on warm modern coasts, and the related spiny cidaroids. But still the reptiles dominated the scene, and above all towered the 45-feet long, 20-feet tall, savage-looking carnivorous tyrannosaurus.

At the end of the Cretaceous these giant reptiles were totally

74 Belemnite shell of Jurassic age; belemnites are distant relatives of the cuttle fish. England.

75 Crinoidal limestone of Jurassic age. To find such perfectly preserved specimens of the crinoid, which rapidly disintegrated when it died, indicates the fossilization of an assemblage almost as it lived. Germany.

51

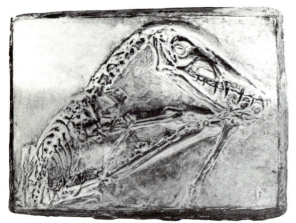

76 A reptile-like bird (*Crassiraspis*) from the Jurassic of Bavaria marks the advance of the early stages in the development of flight.

wiped out. The drastic reduction in sea area may partly account for this, as may the cooling climate and the vanishing swampy low ground but no one of these causes seems to provide an adequate explanation. The demise of the reptiles left the arena clear for the explosive evolution of the mammals in the following sixty million years of the Tertiary and Quaternary.

But we leave that story for a moment to turn to consider the relative position of the continents, for probably sometime in the Jurassic there began to develop the great fractures that progressively widened about an inch or so a year to produce the Atlantic Ocean. The mechanism for this is still in doubt, but new volcanic rocks rising in the middle of the Atlantic have built hundreds of kilometres of new sea floor, with early Tertiary volcanoes on either side of late Tertiary and Cretaceous lavas lying closer still to America and Europe. Enormous quantities of lava were also

erupted on land during this time, and the Tertiary volcanic centres of Britain provide some of the most fascinating ground for observing geological phenomena.

While the gap between the continents in Tertiary times was widening, flat-lying sheets of sand, clay and limestone made of countless millions of the small shells known as nummulites were being deposited over large parts of the United States as well as in the shallow seas that covered much of south-eastern England. In Europe and Asia, great mountains were built at this time —the Alps and the Himalayas and all these Tertiary rocks contain the evidence of the evolution of mammals towards their modern forms. Stages in the growth of the horse from a five-toed form the size of a small dog, and the development of cats and dogs and elephants, bears and cattle are all shown in some detail. But most important of all we find in the latest Tertiary the forms that seem to be the forerunner of modern man, and through the Pleistocene with its intermittent spreads of ice we can find fragments of man's early history.

The evolution of life can be read from Cambrian times until the present, but what of Precambrian times, what signs of life are there in these ancient rocks? Here are signs of the soft jelly-fish, of worm borings and tracks, and limestones show the presence of algae in large reefs. The oldest of these may have lived more than 3,000 million years ago. The Precambrian is full of limestone, and often contains carbon-rich rocks. Much research is being carried out into the chemicals of life in these old rocks. There is much yet to discover.

77 (*left*) The chalk cliffs of Dorset, England,

78 A perfectly preserved million-year-old ginkgo leaf from the clay of the Miocene period of Spokane, Washington.

79 An assortment of prized minerals including fluorite, barytes and zinc blende.

THE NAME mineral is commonly given to substances of economic importance, but to the collector all the crystalline matter of the earth is mineral. All rocks are made up of mineral grains or crystals and, indeed, only a dozen or so substances are required to build the overwhelming majority of rocks. There are however thousands of more rare and often exotically coloured substances.

80 Typical cubic crystals of microline feldspar. Colorado.

81 Natural opal – the precious variety known as landscape opal. Queensland, Australia.

One of the great joys of fieldwork is to find ever better examples of delicate crystal aggregates and robust fiery masses.

At first the task of identifying the newly discovered sample seems disenchanting, but the necessary detailed examination and hunting through the available colourful texts and catalogues adds to the pleasure of the find. It may take the finder into the splendid collections of museums or it may put him in touch with other collectors and experts who are always willing to help and to provide information on good collecting areas. As knowledge grows so does the fascination of collecting.

The first stage of identification is to record as many of the properties of the mineral as possible—whether it is transparent or opaque, whether it is strongly coloured or perhaps variably coloured. It may have some characteristic shape or structure, it may be lustrous, pearly, resinous or earthy, heavy for its size or light. Its hardness can be estimated and the way it breaks may be informative. With practice, these properties can be recorded at a glance and the identity immediately determined. It is the external crystal shape that makes the best starting point, for all minerals that form good crystals show different external forms. The flat-surfaced crystal form results from the internal make-up of the crystal. The whole is made up of atoms that are arranged in a regular three-dimensional pattern of rows and layers. Models of such structures are often exhibited in museums; by using coloured balls to represent the atoms the displays are made extraordinarily decorative as well as instructive.

The size of the atoms that go to build the diverse crystals, as well as their pattern of arrangement, vary so that no two structures are exactly alike. The crystal faces or facets are parallel to the planes of atoms in the crystal and hence directly reflect its internal nature. There are in fact only a few broad crystal types and the collector is soon able to recognise the groups to which a sample belongs. In order to go further it is necessary to measure the angles between the flat surfaces—a task that may not be very easy if the crystals are fragile or very small or much intermeshed. One purpose of making these measurements is to find the characterizing symmetry of the crystal, for it is this symmetry that allows the crystal to be placed in its appropriate group. It is often possible, however, to judge the symmetry of good crystals without measurement—to recognize the six-sided forms of the hexagonal crystals such as those of apatite or beryl or the threefold arrangements of faces of the trigonal quartz, calcite or tourmaline. The faceted brick-shaped orthorhombic crystals exemplified by topaz, barytes or olivine [figure 83] contrast with the faceted bricks of square cross-section of tetragonal crystals such as zircon or cassi-

55

82 Crystals of calcite known as butterfly twins Lancashire, England.

terite. Perfect cubes such as those of galena are sometimes easily seen to have exactly the same symmetry as fluorite or garnet. Monoclinic crystals have some of their faces inclined to the remainder as in gypsum or orthoclase feldspar, whereas triclinic forms such as microcline have all their faces inclined to one another [figure 28].

Some minerals have a crystal structure on such a small scale that it cannot be recognised as such. These substances are described as cryptocrystalline and include banded agates or the green layered copper mineral called malachite. Opal, on the other hand, is truly without crystallinity and is termed amorphous. It is frequently formed by the replacement of some other substance; Australian opals, for example, are defined by replacement of fossils with the hydrated silica of which opal is made.

Crystals may sometimes grow in two directions at once—as though two crystals were developing—or several parallel or intersecting growths may occur. This is termed twinning and adds much to the complexity of facial pattern shown by mineral aggregates. The swallow-tailed twin of calcite demonstrates this phenomenon.

The physical properties of minerals also depend on their atomic make-up, and we find that they sometimes split readily in directions parallel to crystal faces. Evidently the atoms are not held tightly together across these planes of weakness or—technically—planes of cleavage. Excellent cleavage is shown by the flake of micas.

The hardness of minerals also reflects their internal structure—substances such as diamond in which the atoms are extremely tightly bound together are exceptionally hard, while the easily cleaved talc is so soft that it can be scratched with the fingernail. By placing minerals in their order of hardness we make a list which can be most useful for identification purposes. The list is compiled in ten steps from talc to diamond, and several common or garden substances can be used to make some of these steps. On this scale, a fingernail, for example, has a hardness of 2, a bronze coin is 3, a pocket knife, 5 and glass 6.

Today we commonly employ X-ray machines and make chemical analyses to identify minerals, and there is almost no end to the information that can be found about each substance. It is equally true that the experienced collector will be able to identify his material, new and old, simply by looking.

83 Crystals of olivine.

Copper, Silver, Gold

IN 1848 Samuel Brannon announced in excited and dramatic style his find of gold from the America River, California, and started the greatest gold rush of all time. But he was following in a trail of footsteps that stretches from the earliest history of man. The bright yellow glint of the metal is so readily spotted that it is easy to understand how neolithic man came to treasure it—the mineral of the Sun Gods. The exquisite mouldings, carvings and engravings of the Egyptians around 1350 BC have been found in the Pyramids, and the inner coffin of Tutankhamen—the boy king—weighed more than two hundredweights. As far back as the fourth millennium BC, worked gold appeared in the Nile Valley. Gold has also of course been one of the main objectives of plunderers and warmakers. The Spaniards robbed the Incas and Aztecs, Alexander the Great won large quantities from Egypt, and much of the Roman glory was based on Grecian gold. So the metal continues to be circulated and recirculated and the natural supply

84 Gold-prospectors in California in the mid-nineteenth century, using the 'long tom' to sift alluvial gold.

of newly dug gold fails to meet man's apparent need.

Gold is easy to recognize by its softness, density, malleability and chemical resistance. The minerals most like it are pyrite, which is much harder and makes gleaming crystals often well worth collecting in their own right, and the slightly greenish-yellow copper pyrite. The latter is soft, but when scratched it yields dull greenish powder. Gold occurs as heavy rounded nuggets or as tiny flakes or specks in the cracks and hollows below the sands of river beds, and the river extracts the metal, from decomposed veins or 'gossans' upstream of the alluvial deposits, from what is known as the mother lode. Gold is also found in volcanic hot springs and much gold seems to have been formed in veins when certain granites rose into the crust. The Californian mother lode is of this kind. But the world's richest accumulation appears to be the metamorphosed river gravels of Witwatersrand in South Africa. These have been mined to such deep levels that the heat of the mines has become almost intolerable, reaching some 120° Fahrenheit. A recent boring has located a possible extension of this conglomerate in Orange Free State, and this bed is already yielding gold from a depth of 5,000 feet.

Gold in Britain is rarely found, though a few individuals have occasionally eked out a living by panning in some of the rivers of Scotland. A nugget of nearly four ounces has been found in Sutherland, and a few grains of alluvial gold are to be found in Cornwall. In North Wales gold has been obtained from quartz veins in the Cambrian slates and the metal taken from this locality has been used for royal jewellery. A particularly attractive and unusual variety, consisting of plates and cubes of crystalline gold, was found a century ago in Nag-Yag in eastern Europe.

As gold epitomizes the sun, so silver has been associated with the moon and moongods. At one time, in early Egyptian dynasties, the metal was more valuable and thought to be scarcer than gold. It is lighter, size for size, than gold and much more reactive with common chemicals. But in spite of its reactivity, silver is found in the pure state in nature. These natural occurrences are usually of contorted, twisted wire-like masses or delicate skeletal crystals. Much more commonly, silver occurs in the form of one of its compounds and often is but an impurity in a mineral that would otherwise be almost worthless. The lead sulphide called galena crystallizes, when freshly broken, into bright cubes, and tarnishes very slowly. The galena containing a high percentage of silver is usually a dull grey. Commonly one mineral will indicate the almost certain presence of another, and finding niccolite or other nickel or cobalt arsenic compounds is a sure sign that silver is around.

85 World's largest gold vein. Kenya.

Copper is found also as a native metal [figure 33], and much more common, of course, are its numerous compounds, which are among the most colourful of minerals. The most important accumulations of metallic copper are in Bolivia and the Keweenaw Peninsula of Lake Superior. Basalts rarely contain crystals of the metal, and these are curiosities rather than useful ores. It was the metallic deposits that were first used by man at least 8,000 years ago. The early users probably worked the copper by cold hammering and it was much later before the process of annealing was discovered; this certainly pre-dated the technique of smelting compounds of copper to extract the metal. These green, blue and iridescent ores are much in use for decorative purposes, and are to be found in countless localities throughout the world.

One of the most colourful of copper minerals is malachite [figure 88]. This mineral occurs in intersecting spheres that are built up of shell upon shell of concentric zones of light and dark emerald green. English railway engineers assisting in track-laying in the USSR in the nineteenth century occasionally returned home with large masses of malachite up to three feet in diameter. Not all of it was of good quality, but much malachite is good enough to work into thin, sometimes curved, pieces for decorative mounts or veneers, and some Russian craftsmen used the material for table tops and urns. Malachite is a most important ore of copper and forms the basis of the great copper belts of Katanga, Rhodesia and Zambia. These belts are formed by the alteration of other copper minerals by weathering. They are therefore essentially superficial ores, and it is this surface development that produces the regular banded structure. The mineral is a basic copper carbonate; another very closely related copper carbonate is the brilliant blue crystalline azurite. This is so little different from malachite that the two often occur together. Malachite often assumes the crystal shape of azurite–replacing the azurite with but the slightest change of composition. The brilliant blue chalcanthite is water-soluble and therefore rare.

A fine deep green mineral found in the Congo is dioptase [figure 92] which is sometimes called emerald copper. It was originally found in the Urals and frequently makes delightful pea-sized crystals, which are sometimes twinned and have occasionally been cut as gemstones. The crystals may sometimes occur separately or in aggregates on a matrix of other minerals. Many strikingly coloured specimens of dioptase are nodular crystalline aggregates which are very much like the massive form of azurite known as chessylite. Chessylite is named from its occurrence at Chessy in France and very large specimens can also be obtained from the Congo. Remarkably fragile balls of pale blue chessylite obtained from Australia can be crushed to make a fine blue pigment for painting on vellum.

Cornish mines have yielded moss-like native copper, malachite and azurite often containing or edged with such minerals as the reddish copper oxide, cuprite. Cuprite is often well crystallized though the crystals rarely exceed half an inch across. In the same assemblage are chalcocite, the iridescent peacock ore known as bornite or erubescite [figure 93], and the yellow-bronze tetrahedrite sometimes makes excellent crystals, the shape of which give the mineral its name. Another blue or green copper mineral is called chrysocolla; its name tells us of a former use of the substance, for it means 'gold glue' and was used in soldering gold.

86 Loading the 'hopper'. As the gold is heavier than any dust or fragments of rock, it is left on the lowest board.

Blue John, Amethyst and Agate

MANY OF the valuable ore minerals are found in veins together with crystalline minerals that, though quite useless as ores, are often more attractive than the common commercial minerals. One of the most colourful of such minerals is called fluorite or sometimes fluorspar and the occurrence of this mineral in the north of England exemplifies the similar vein associations of the world. Here the fluorite occurs in veins and line cavities, and is accompanied by lead and zinc sulphide ores. In Derbyshire, the lustrous grey mineral mined for lead is called galena, and there is also umber (used in paints). There is a great variation in the colour and quality of fluorite. Some is white and often contains stringers of golden pyrite within it, giving the appearance of yellow dust that has settled on the faces of the crystals as they grew. Other fluorite is clouded with yellow barytes which also forms thin bladed crystals between the pyrite-flecked crystals. Sometimes whole veins may be of poorly coloured small crystals, but here and there is usually to be found exquisite blue or pink and sometimes green or purple fluorite. Some of this purple fluorite is of sufficiently good quality for use in the highly magnifying lenses of expensive microscopes.

87 A fine water-clear quartz crystal aggregate intergrown with lustrous black tourmaline. Brazil.

High-quality fluorite is also found in France and in Illinois and Kentucky, and here some limestone caves contain enormous single cubic crystals of the mineral. Fine pink crystals are found in Switzerland, and a well known fluorite is that called Blue John which is mined in Derbyshire. This is a fluorite made of crystals set close together and all are banded across with purple and various shades of blue. This material was carved by the Romans into bowls and vases and many other forms. The faces of many fluorite crystals have a thin coating of crystals of quartz

89 Crystallized malachite.

and calcite. Some crystal masses are very beautiful in colour and almost iridescent when they are freshly mined. A few weeks later, however, they have lost this rare beauty and become grey and dull.

Superb sets of gemstones cut from the various coloured fluorites can be seen in many national museums. A difficult colour to obtain is a good yellow, and there is only one area in England that has yielded fine, clear, yellow stones. Purplish, bluish-purple, green and clear fluorspar is not as difficult to find, but little of it is suitable for cutting, for the relatively easy cleavage of fluorite causes a high percentage of breaks when the mineral is polished. A group of large, dark-purple crystals, very clear and almost flawless, forms the base of one of a pair of fine altar candlesticks to be seen in Westminster Abbey. The fluorite crystals, one slightly interpenetrating the other, are about six inches across the faces. The other candlestick of the pair has a base consisting of a very fine and rare quartz crystal group mined in about 1820 at Tintagel, Cornwall.

From the fluorites and associated minerals we turn to minerals of similar colour which are frequently mistaken for them—the amethysts or coloured crystals of quartz. In fact, the two groups of minerals differ noticeably in crystalline form and hardness. Fluorites have cubic crystals with a hardness of 4. Amethysts on the other hand make crystals belonging to the trigonal system and will scratch glass. The colour of amethyst may be caused by traces of manganese in the mineral.

Amethyst crystals are often found as a lining in cavities [figure 96]. They are attached to a thin wall of banded agate which is usually grey, green or greenish-brown. Brazil and Uruguay are constant sources of specimens. The shade of the purple amethyst crystals varies. Normally the crystals are pale, but in mass they naturally look darker, and of extreme beauty are the very dark aggregates of crystals that line some egg-shaped cavities. Most amethyst crystals are quite small and up to a finger thick, but sometimes large amethyst crystals like giant hands have been discovered. They are usually short-sided and terminate in pyramids that make the crystals sparkle. Occasionally there are small areas of white or milky quartz in the cavities too. The faceted amethyst gemstones are cut from broken masses of crystals, each piece of broken crystal being sorted for colour against a bright light.

There are quartz crystals of various attractive colours from many countries. Smoky-quartz crystals come from Switzerland and South America, and the very dark material used for cutting may be from Spain. The large, broken masses of yellow quartz called citrine invariably come from Brazil. Rose quartz has been known and used for carving for hundreds of years in India, China and Japan and other far eastern countries. Madagascar, Africa and Brazil are also well-known sources. The depth of colour in rose quartz varies greatly. A specimen needs to be of a certain thickness to lose the white tinge that subdues the pink shade. Rose quartz from some localities has been said to fade over the years. Really black crystals of quartz, which are called morion, are quite rare. They are very symmetrical as a rule and almost perfect in shape.

Quartz crystals, either alone or in groups, are incredibly diverse

90 A rock cavity in which the cryptocrystalline variety of quartz called chalcedony has grown as linked and intersecting spheres.

in shape and formation. Many exquisite white groups are English in origin. White crystals in flat groups and on fluorite make superb room decorations. We can find very short crystals and good pyramids up to an inch in diameter. Some of the groups are very large and can be up to three feet across. In iron mines very clear double-ended pyramids of quartz with little or no hexagonal trunk can often be found. These are often associated with iron minerals and the quartz crystals may contain iron glance. This is similar to the lovely Venus hair of Brazil and Madagascar which consits of golden hair-like filaments of rutile enclosed in clear quartz. In Derbyshire there is an interesting occurrence of small bipyramidal quartz crystals which are known as Derbyshire diamonds, while those from West Cumberland iron mines are called Cumberland diamonds.

Brazil and Madagascar have produced quartz crystals of immense size and many specimens from Brazil show fine inclusions of pyrite crystal cubes and gas bubbles. Some display on the vertical faces rosettes of pink and green micaceous minerals such as anthophyllite and sometimes combinations of green tourmaline and pink rubellite [figure 99]. Tourmaline is also found as inclusions in quartz crystals.

Related to quartz but not showing definite crystals are chalcedony and agate [figure 90]. Agates have well-banded structures and are made up of concentric ovals of variously coloured

65

91 (*below*) The water-soluble blue copper mineral, chalcanthite. Mexico..

92 (*right*) The copper mineral, dioptase, from Katanga, Africa.

93 (*opposite*) The iridescent erubescite or peacock ore. Rhodesia.

chalcedony. Onyx is similar but the chalcedony layers are less visible. The black and white parallel layers of onyx are the result of artificial staining, a simple process known for several thousands of years in which the chalcedony is impregnated with sugar solutions and then soaked in sulphuric acid. The green, yellow and brown varieties are modern examples using synthetic dyes.

Red and yellow jaspers are another variety of cryptocrystalline quartz containing iron. Sometimes the colour is remarkably bright and large jaspers are striking when banded. Some of the finest chalcedonies are shaped like stalactites and are quite translucent, and fine crystal-lined cavities can be found in flints on occasions. The chalcedony geodes of Uruguay [figure 94] have water moving inside them that was forced in under pressure during torrential storms. Other forms of chalcedony include the deep green heliotrope which with flecks of red jasper is called bloodstone. Another variety beloved of the ancient Romans is the leek-green prase. It is found in Italy and in Bohemia where is to be found Europe's oldest lapidary centre, where the sixteenth-century quartz crystal tazzas were ground and polished and garnets cut.

94 Chalcedony containing water, which causes the darkening on the right-hand side of the specimen. Uruguay.

Valuable metals and deadly poisons

THE SOFT, colourful, vein mineral fluorite is often associated with lead and zinc minerals. The ancient lead mines in Yorkshire, England, which were worked by the Romans provide a typical picture of mining in such country. The mineralized zones are in limestone and consist of galena veins up to two feet thick, some encrusted with white crystals of cerussite—a lead mineral. The fluorite often lines cavities in the mineral zones. At one place zinc blende has been mined. Most crystals of zinc blende are pea-sized and arranged in clusters, and occasionally cleaved faces one and two inches across can be seen. The mineral often appears to change colour from a rich deep brown to black. The crystal masses show up strikingly against a white matrix of quartz, which is often in the shape of typical fluorite cubes—the fluorite having been dissolved and replaced by the quartz.

Such lead–zinc veins make wonderful mineral specimens. There are for example the delicate watery-blue cerussites (lead carbonate) from Tasmania and adamantine anglesite (lead sulphate) crystals from Sardinia. Another mineral often found in old collections is minium, a form of red lead from Siberian localities, and in Cumberland, England, is a rare and renowned occurrence of campylite. The orange, barrel-shaped crystals are set in a mass of pyrolusite (a manganese mineral described later). Nearby, green, nail-head-shaped crystals of pyromorphite are found.

The beautiful blue crystals of linarite are a lead-bearing variety of the copper mineral azurite, and lead frequently forms minerals in such combinations with other metals. There are for example the richly coloured vanadium-bearing mineral called vanadinite, the molybdenum-lead ore wulfenite and the hyacinth-red pencil-shaped crystals of lead chromate which is called crocoisite or crocoite [figure 100].

Lead minerals are also found with discrete grains of molybdenum ore, and confusion can often arise between the metallic grey lead ore galena and, the lighter grey ore molybdenite. There is however no possibility of confusing the ores of tungsten with other minerals. Wolfram is an iron and manganese tungstate, is usually grey-black or reddish-brown and is extremely heavy for its size. It is often associated with the calcium tungstate known as scheelite and the latter may occur as clear or white crystals coating the wolfram. Miners used to fake specimens by gumming scheelite crystals into crevices between quartzes at the ends of bladed wolfram crystals, hiding the forgery with flaky mica plates.

95 Cubes of fluorite. England.

The manganese in wolfram does not need to be extracted, for there are countless valuable manganese ores throughout the world. Psilomelane is particularly well known and forms some most unusual structures. Sometimes it forms shining black balls the size of ping-pong balls, fastened together by rods, and at other times it occurs in bubbly, slag-like dark brown masses. These structures can be found practically every where manganese is mined, in Africa, America and Japan for example. A very beautiful and spectacular manganese ore is massive rhodonite [figure 97]. This is a hard variety very much like jasper but of a paler, rose-pink shade.

96 A geode with fine amethyst crystals. Brazil.

97 (*left*) The pink silicate of manganese known as rhodonite forms delicate radiating aggregates. Devon, England.

Specimens can be found in many parts of central and eastern Europe, and Transylvania was a famous old mining area for rhodonite in medieval times. Lovely banded specimens of the rose-pink manganese spar or rhodochroisite have become quite common since the 1940s [figure 103]. Great masses have been mined in South America, particularly in Brazil. Many specimens have been carved into ashtrays to expose more fully the stalactitic and agate-like veins, loops and curves of the many-coloured bands. Fowlerite is a rarer manganese mineral found with zinc minerals at Franklin in New Jersey.

The limestone hosts to the fluorite and lead-zinc veins often contain large masses of a colourful variety of iron ores. These ores are frequently associated with other crystals such as barytes [figure 101] which may be water-blue or white and tinged with red. Also there are masses of exceedingly beautiful clusters of 'dog-tooth' and 'nail-head' calcite [figure 104] often tinged with red iron oxide at the tips. The iron-ore mines produce haematite, which makes knobbly red-black kidney-shaped masses [figure 102]. Specially long, triangular 'nails' of haematite, are valued very highly when cut and polished into black faceted ring stones. Most of the iron ore of the United States appears in the form of haematite, and enormous deposits of this mineral stain the earth blood-red in Minnesota. Mining in South America has brought to light gigantic crystals of the deep blue or green vivianite or iron phosphate, and associated minerals which were formerly thought to be very rare. When such minerals were first found at Potosí, in Bolivia, they were much sought after by collectors

98 Brown slag-like masses of the valuable manganese ore, psilomelane. Cornwall, England.

101 Water-clear crystals of the barium mineral, barytes. Cumberland, England.

because of their great beauty and exquisite colour. Early specimens were quite small, but these days large blades of vivianite a foot or more in length are commonplace. They are a dull blue, but they may well have been clear before they were mined and exposed to the air. Variscite is an attractive phosphate mineral containing aluminium [figure 106].

Structures with much the same banding as haematite may be made of limonite. This iron ore is deep chocolate brown and the layering is usually marked by yellowish bands. The limonite may be a replacement or alteration of the haematite, and golden cubes of pyrite are frequently replaced by this earthy mineral. The very well-known semi-precious stone called marcasite is a much less common mineral.

Along with iron minerals there are sometimes to be found the ores of nickel and cobalt, but these metals are more often found in some of the great gabbros of the world. Mines in Sudbury, Ontario, have produced smaltite and kupfernickel for many years, together with other rarer minerals that occur in banded veins. Smaltite is essentially an arsenic compound of cobalt, but is found associated with chloanthite, an arsenic compound of nickel. Kupfernickel (also called copper nickel) is also an arsenide of nickel. It, too, often contains a little cobalt. Specimens of smaltite and kupfernickel generally have to be polished before they can be exhibited. Cobaltine-calcite is a fine red [figure 107]. New localities for these minerals are being worked in many parts of the world, but as yet no spectacular specimens have been unearthed. Garnierite is a bright green, nickel-bearing mineral of very attractive appearance which is worked in New Caledonia. The immensely poisonous arsenic is widespread as a mineral and is frequently found in veins associated with granites as well as in the gabbros. The vivid red realgars and massive yellow orpiment [figure 109] – the arsenic sulphides – are widely found and localities in central Europe were known to early collectors. Around 1920 the prettiest of specimens came from Nevada, and during the Second World War arsenic minerals were mined in Australia. Special laws were passed to prevent miners contracting arsenic poisoning through the skin by overworking and perspiring. Borneo and Sarawak are other areas in which arsenic minerals can be found.

Nearly as poisonous as arsenic is antimony which at one time was much used in cosmetics and today is frequently used on the boxes of safety matches. Indeed the simplest test for identifying stibnite, which is antimony sulphide, is to strike a safety match on the crystals. The great silver-grey knife-blade crystals of stibnite are the star pieces of many collections and a vast number of

102 Lustrous red-black intersecting spheres of haematite. Cumberland.

such samples have originated from Brazilian and Japanese mines. Many fine-grained specimens from Eastern Europe are superficially altered to become coated with the yellowish-white cervantite. Another of the natural poisons is mercury, and globules of this valuable and brilliant substance are sometimes found in veins made of the deep red fibrous mercury ore called cinnabar. The occurrence of radioactive minerals also might present something of a health hazard, and the collector would be well advised to store his uranium-bearing specimens in separate containers. The recent demand made prospecting for these minerals as stimulating as the great gold rushes. Now, however, sufficient is known to have taken much of the steam out of the searches. The minerals are often extremely attractive and include autunite and torbernite which form delicate green and yellow scaly crystals on veins in granite, and some fluoresce in ultra-violet light.

Some rock-making minerals

105 A group of perfect glass-clear quartz crystals. Madagascar.

THE VAST majority of rocks are combinations of but two or three out of a dozen or so common minerals, and occasionally these few substances are sufficiently well grown to be worth collecting as mineral samples. Most of these minerals are, in turn, made of a selection of only a few elements, and the most common atoms are oxygen and silicon. These two atoms are set together to make tetrahedral building bricks of four oxygens around one silicon, and the bricks are piled together with a few other atoms such as sodium, potassium, magnesium and iron to make the so-called silicate minerals. The widespread clear, hard, glassy mineral quartz is made wholly of the tetrahedra. Its clarity is responsible for its name, which means clear ice.

The feldspars are built of tetrahedra linked together to make a three-dimensional network within which are scattered sodium, potassium and calcium atoms. The pink feldspars of granite are for example potassium and sodium silicates, and the feldspars of gabbros are sodium and calcium silicates. In the micas the tetrahedral bricks are linked into layers: the layers are joined by atoms such as aluminium and iron and spaces in the structure are occupied by potassium and magnesium. As a general rule, transparent, pale-coloured minerals that do not scatter light very strongly are often frameworks of tetrahedra containing sodium, potassium and calcium, whereas those silicates containing iron, chromium, manganese and other heavy atoms tend to be strongly coloured and scatter light brilliantly.

Garnets fall into the latter category. These occur in a great variety of rocks and are especially important in schists and marbles. Usually they make small grains, but sometimes perfect golf-ball-sized crystals are found and occasionally, as at Gove Mountain, New York, they may be a foot or so in diameter. Garnets are quite hard; poor-quality material is used as an abrasive, but clear crystalline garnet makes superb gemstones. They occur in a great variety of colours. From the Vilni River in Siberia comes pale green calcium–aluminium garnet called grossularite [figure 109]. Mexico is a source of clear pink grossularite, and the yellow cinnamon stone or essonite found in Maine and Ceylon is also grossularite. Magnesium–aluminium garnet is the clear brilliant crimson pyrope so much prized in the nineteenth century. Fine specimens of this come from Bohemia. A close relative of pyrope is almandine which is the iron–aluminium garnet. Its deep red colour and great hardness has attracted the jewellers of countless generations, and it was known to Pliny. Manganese garnet, which shows a variety of colours, is called spessartite. From Madagascar comes clear orange spessartite, and the European crystals are normally reddish brown. Jet black melanite is calcium–iron garnet and is common in some igneous rocks, and polyadelphite is a brownish-yellow relative of melanite from Franklin, New Jersey. Another relative is the bright green demantoid gem garnet found in Val Malenco in Italy.

Another common silicate mineral is the iron-magnesium-bearing olivine, or chrysolite. The clear green variety is used as a gemstone and called peridot. In ancient times the people of the Middle East used peridots extensively in their jewellery designs. Most of their specimens probably came from the so-called Levant or the Red Sea area. Peridot is found in basic volcanic rocks, in serpentines and in altered magnesium-rich rocks. Tremendous force is required to extract the olivine nodules from which the

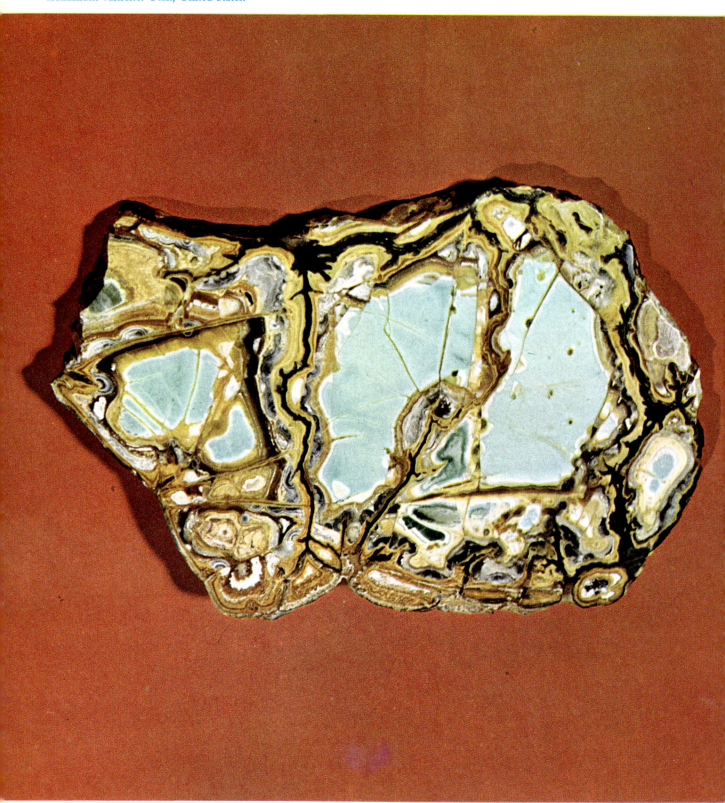

gems are cut. The Rhine lavas and various basalts contain nodules up to two inches in diameter. Some rocks such as the dunites of the Carolinas are made up almost entirely of olivine.

The silicate group called pyroxenes are also coloured calcium, iron and magnesium minerals, and these dark crystals form the dark part of gabbros. They are most easily recognized by their two sets of cleavage planes set nearly at right angles to one another. Occasionally these minerals are of gem quality, and a rare white or lilac-coloured pyroxene called spodumene provides us with the gemstone known as kunzite. It was first discovered at San Diego in California and then in Madagascar, and now many fine crystals come from Brazil.

Another group of minerals related to the pyroxenes in composition and structure is the amphiboles. These also possess two sets of cleavage planes, but this time the cleavages are at about 60 degrees to one another. The crystals are lustrous black, brown or green prisms; a very common variety of amphibole is the very dark hornblende.

The popular gemstone zircon is found in a great variety of rocks but is most spectacularly developed in the pegmatites—the concentrates of the residual liquid of solidifying granite. Zircon is similar in structure to olivine. Black, pink and green tourmalines and beautiful aquamarine beryl crystals are other common coloured silicates found in pegmatites, along with the usual quartz and feldspar crystals.

Topaz is another superb example of a finely crystallized pegmatite mineral [figure 108]. Specimens from Brazil are usually orange-coloured and up to an inch across. White and clear topaz are found in Nigeria and Japan. The finest specimens are eight-sided crystals several inches across. A rare blue topaz is occasionally found in Brazil and Russia. Topaz crystals are seldom found together in a matrix because they fall out easily. Most of the big specimen discoveries have occurred since Nigeria has been prospected for tin; stream-rounded boulders of water-clear topaz up to a foot in diameter have been found. Rubies and sapphires are varieties of the extremely hard aluminium oxide mineral known as corundum. Even the non-gem quality corundum is rare and is produced by very high-grade metamorphism in the vicinity of some basaltic intrusions.

Making crystals glow

Some minerals possess the property of glowing when ultra-violet light is shone on them. This property is called fluorescence, after fluorite which shows it beautifully. Ultra-violet lamps and other equipment are now quite cheap and can be set up anywhere and take up little space. The bluish light of the fluorite found in the north of England is very powerful and it is curious that fluorites from other localities do not fluoresce to such an extent.

Several other minerals besides fluorite exhibit fluorescence. A few of them are mentioned below. But it is always worthwhile putting a new mineral find under ultra-violet light. A group of fluorescent minerals arranged together make a fascinating display.

Most calcites do not fluoresce, but those that do emit a beautiful soft pink, but short-range, glow. Aragonites which come from Sicily fluoresce a similar colour if the ultra-violet lamp is close to them, but those from other localities do not fluoresce.

Some Australian opals fluoresce white, willemite from New Jersey gives a lovely strong green colour and the radioactive autunite, sometimes found as a thin film on granite, also fluoresces a deep green. True rubies glow a brilliant red in ultra-violet light as do clear spinel rubies. The red fluorescence of hackmannite is more of a velvety red. Yellow Canadian scapolite fluoresces a bright yellow and haüyne, which is found in small blobs and crystals in some lavas, exhibits a bright yellowish-red fluorescence.

108 Natural crystals of the semi-precious topaz from Brazil and Nigeria.

110 Pale green garnets – grossularites – from Mexico.

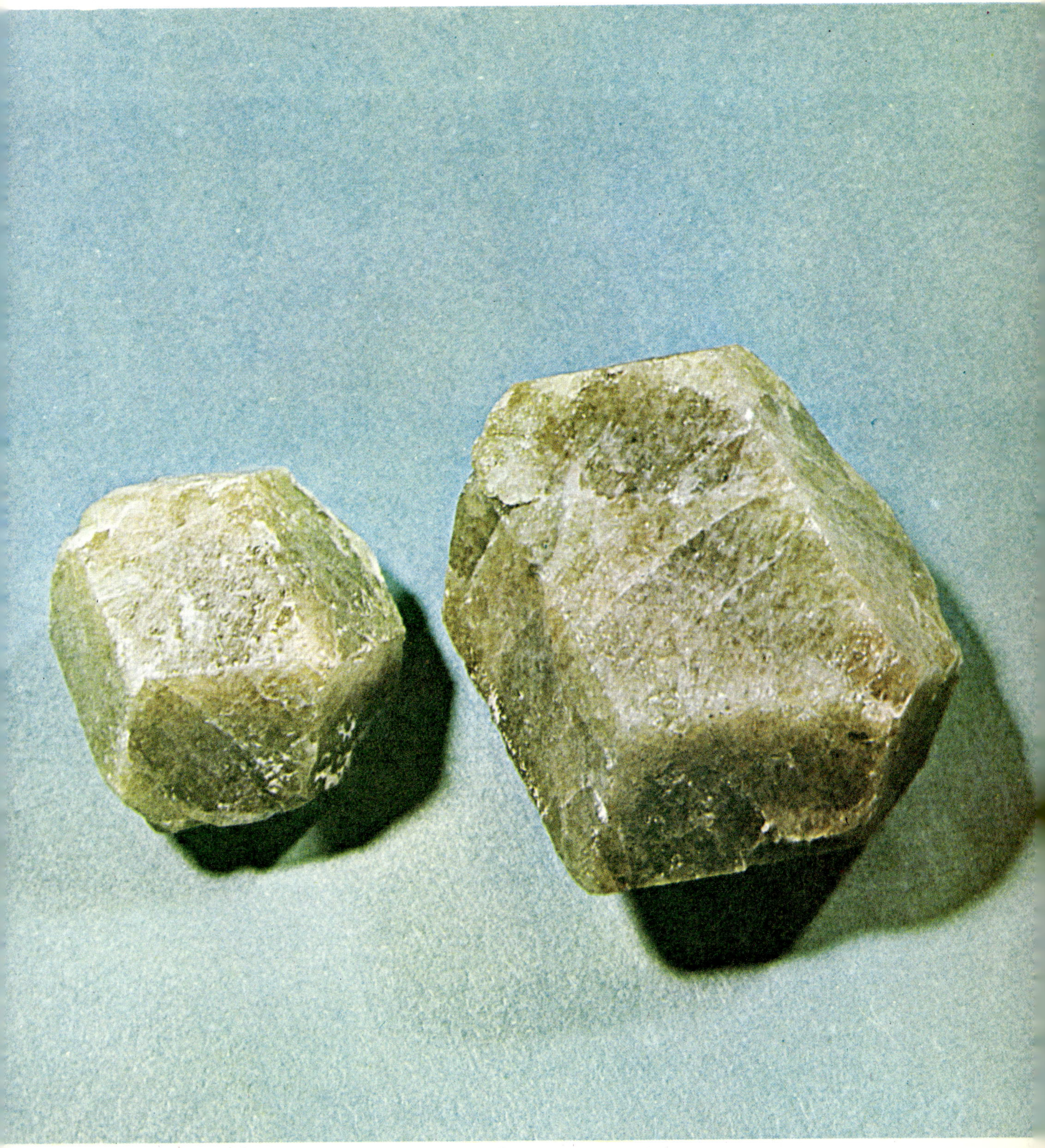

111 Basalt columns of the Giant's Causeway, Co. Antrim Ireland.

Eras	Period and system names	Subsystem	Length of periods etc in million years	Age to start of eras
Cainozoic	Quaternary	Holocene-recent	2-3	
		Pleistocene-glacial		
	Tertiary	Pliocene	9-10	
		Miocene	13	
		Oligocene	16	
		Eocene	20	
		Paleocene	12	70
Mesozoic	Cretaceous		65	
	Jurassic		45	
	Triassic		45	225
Paleozoic	Permian		45	
	Carboniferous		80	
	Devonian		50	
	Silurian		40	
	Ordovician		60	
	Cambrian		100	600

Proterozoic = Cryptozoic = Precambrian Era

Oldest known rock is approximately 3,500 million years.

Origin of crust is about 4,500 million years.

GLOSSARY

Agate	banded chalcedony
Agglomerate	volcanic rock made of large fragments of rock and lava 'drops'
Ammonites	fossils composed mainly of spiral segmented shells
Amygdale	vesicle infilled with later mineral growth
Andalusite	orthorhombic crystals of aluminium silicate found in medium- to high-grade metamorphic rocks
Andesite	fine-grained volcanic rock lacking quartz but with feldspar, pyroxene or horn-blende
Anhydrite	salt mineral, water-free calcium sulphate
Antimony	an element
Apatite	a calcium phosphate making hard hexagonal crystals
Aplite	medium-grained granitic rock found in veins and patches
Apophyllite	calcium, potassium zeolite forming prismatic crystals
Archeopterix	early fossil bird
Azurite	a blue crystalline copper mineral
Barytes	the principal ore of barium, makes good orthorhombic crystals
Basalt	fine-grained dark volcanic rock made of feldspar and pyroxene
Batholith	a mass of granite measuring several miles or more in diameter and often with sharp, steeply inclined, margins
Belemnites	cone-shaped skeletons of relatives of the squid
Beryl	green hexagonal silicate crystals
Bornite	peacock ore – a colourful copper mineral
Borolonite	a syenite with white spots and melanite garnet
Calcite	crystalline calcium carbonite which occurs in trigonal crystals
Carbonatite	a coarse igneous rock rich in calcite and other carbonates
Carborundum	an artificially produced abrasive
Carnallite	a salt mineral, ore of potassium
Cassiterite	a tin ore
Celestite	an ore mineral of strontian
Cerussite	an ore of lead
Chabazite	sodium calcium zeolite making trigonal crystals
Chalcedony	cryptocrystalline silica
Chalcocite	a copper sulphide – an important ore
Chessylite	a synonym for azurite
Chiastolite	a variety of andalusite
Chrysocolla	a blue/green copper silicate
Ciradoid	relative of the modern sea urchin
Cinnabar	a fibrous red ore of mercury
Citrine	yellow gemstone of quartz
Coal balls	lumps in coal in which the original plants are preserved by petrification in limestone
Cordierite	an orthorhombic magnesium silicate found in metamorphic rocks
Crinoid	a relative of the sea urchin which consists of a cup or head and a jointed stem by which it is attached to the sea floor
Cryptocrystalline	crystals that can not be seen even with a powerful optical microscope

CRYSTALLITE	exceedingly small crystals
CUPRITE	copper oxide – a copper mineral
DETRITUS	disintegrated material of rocks
DIOPTASE	a green copper mineral
DIORITE	a coarse igneous rock made of hornblende and white feldspar
DOLERITE	an igneous rock like basalt but coarser in grain-size
DREIKANTER	pebbles with three facets shaped by sand blasting
DYKES	nearly vertical wall-like bodies of igneous rock cutting across other rocks
ECLOGITE	a metamorphosed gabbro containing garnet and no feldspar
EPIDIORITE	a metamorphosed dolerite or gabbro
EPIDOTE	green calcium-rich silicate mineral making monoclinic crystals
ERUBESCITE	a synonym for bornite
FELDSPAR	a group of common rock-forming silicate minerals, light-coloured and containing sodium and potassium or sodium and calcium
FLUORITE	ore of fluorine making good colourful cubic crystals
FOYAITE	like ijolite, but containing some sodium and potassium – rich feldspar
GABBRO	an igneous rock of basaltic composition but very coarse grain-size
GALENA	a soft grey lead ore
GARNET	very hard variably coloured silicates of iron, calcium, aluminium etc
GEODE	a rock containing an egg-shaped cavity in which fine crystals have often grown
GLAUCOPHANE	a blue or purple amphibole found in metamorphic rocks
GNEISS	coarse-grained rock in which bands of light granular rock alternate with mica-rich schist layers
GRANODIORITE	coarse-grained igneous rock made of quartz, sodium-calcium feldspar
GYPSUM	hydrated calcium sulphate, white monoclinic crystals
HALITE	natural crystals of common salt
HEULANDITE	sodium calcium zeolite making monoclinic crystals
ICHTHYOSAUR	an extinct marine reptile
IDOCRASE	a calcium-rich tetragonal silicate mineral found in marbles
IGNIMBRITE	volcanic tuff formed mainly of glass shards
IJOLITE	a syenitic igneous rock composed of light sodium-rich minerals and sodium-rich pyroxene
IRON GLANCE	synonym for specularite
JADEITE	the mineral of jade, a green metamorphic mineral
KENTALLENITE	a coarse igneous rock made of olivine, pyroxene, mica, potassium sodium and calcium feldspars
KUPFERNICKEL	a nickel mineral
KYANITE	a high pressure aluminium silicate
LAMPROPHYRE	rocks typified by large dark crystals in a background of finer dark and light crystals
LATHS	elongated crystals with rectangular cross-sections
LEUCITE	a potassium silicate making cubic crystals
LEUCITE-TEPHRITE	a basaltic volcanic rock with leucite crystals
LINGULA	shell-fish of Cambrian to recent period
LINGULELLA	shell-fish of Cambrian period
MALACHITE	a green copper mineral usually built in parallel, approximately spherical layers
MATRIX	mass of rock enclosing gems
METAGABBRO	a metamorphosed gabbro
MICA	a common silicate in which the atoms are arranged in layers

MICROCLINE	a triclinic alkali-rich feldspar
MICROCRYSTALS	very small crystals
MINIUM	a lead mineral
MONOCLINIC	one of the seven crystal systems, the crystal faces being referable to three directions, two of which meet obliquely while the third stands at right angles to the other two
MORION	black quartz
MOTHERLODE	the principal mineral vein of a given area
MUGEARITE	volcanic rock allied to basalt but with some potassium feldspar
MYLONITE	a very fine-grained or glassy rock produced by the frictional heat produced by movement along faults
NATROLITE	sodium zeolite making needle-like crystals
NAUTILUS	a spiral shelled form in which the shell is segmented
NICCOLITE	a nickel mineral
NIOBIUM	a rare element
NOSEAN	sodium silicate making cubic crystals
NUMULITES	microfossils with flat coiled shells
OBSIDIAN	variety of volcanic glass
OLIVINE	a magnesium-iron, very dense silicate, often green
ONYX	agate with parallel planar black and white bands
OOLITE	a limestone composed of small pellets each made of concentric layers of calcium carbonate
OPHICALCITE	a marble streaked with yellow and green minerals
ORPIMENT	a yellow arsenic ore
ORTHOCLASE	a monoclinic alkali-rich feldspar
PEGMATITE	very coarse-grained igneous rock
PERIDOT	gem-quality olivine
PERIDOTITE	a coarse rock composed of pyroxene and olivine
PERMIAN	one of the geological periods
PETROLOGY	the study of rocks
PHONOLITE	volcanic rock composed predominantly of potassium feldspar and pyroxene
PHYLLITE	a metamorphic rock in which greenish micas impart a sheen to planes of easy parting
PICRITE	a dark, olivine-rich, igneous rock allied to dolerite
PLEISTOCENE	part of the quaternary system
PYRITE	a brassy yellow crystalline iron sulphide
PYROXENE	a group of common rock-forming silicate minerals, dark in colour and containing various amounts of iron, magnesium and calcium
QUARTZ	clear hard glass-like trigonal crystals of pure silica
REALGAR	a red arsenic ore
RHYOLITE	fine-grained volcanic rock composed of quartz and feldspar, often with much glass
RUTILE	a titanium mineral
SCHIST	medium or coarse-grained metamorphic rock typified by parallelism of micas
SCOLECITE	calcium zeolite making needle-like monoclinic crystals
SELENITE	water-clear crystals of gypsum used for optical purposes
SERPENTINITE	a soft rock often made by alteration of peridotite to give soft flaky layered silicates
SHALE	a very fine-grained (clay) sediment with lamination

SHARD	broken fragment usually of glass
SILICA	silicon dioxide, the material of which quartz is made
SILLIMANITE	high-temperature aluminium silicate
SILLS	sheets of igneous rock parallel to layering of the older rocks enclosing them
SLURRY	free-flowing mud
SPECULARITE	crystalline iron oxide in which the crystals are usually dark and remarkably lustrous
SPHERULITES	ball-shaped pellets composed of radiating crystallites
SPILITE	volcanic rock allied to basalt but much altered
STAUROLITE	an orthorhombic iron aluminium silicate found in metamorphic rocks
STILBITE	sodium calcium zeolite making monoclinic crystals
SYENITE	a coarse igneous rock made of dark minerals and pink potassium and sodium feldspars
SYLVINE	a cubic salt mineral, principle ore of potassium
TETRAHEDRITE	a copper mineral that makes crystals of tetrahedral shape
TOPAZ	yellow or colourless mineral made of aluminium silicate forming orthorhombic crystals
TOURMALINE	coloured silicate containing boron
TRACHYTE	a volcanic rock of sodium and potassic feldspars with a little pyroxene
TRAVERTINE	concretionary calcium carbonate of growths such as stalagmites and stalactites
TRIASSIC	one of the geological periods
TRICLINIC	one of the seven crystal systems – the faces are referred to three directions which all intersect at angles other than 90°
TRILOBITE	a primitive extinct crustacean
TROCTOLITE	a relative of gabbro in which olivine is overwhelmingly dominant over pyroxene
TUFF	a fine-grained volcanic rock made up of bits of crystal, rock or glass
TUNGSTEN	a lead mineral
TYRANNOSAURUS	an extinct giant carnivorous reptile
VESICLE	gas-filled cavity in volcanic rock
VOGESITE	a variety of lamprophyre with hornblende or pyroxene and potassium feldspar
WOLFRAM	a tungsten mineral
XENOLITHS	pieces of pre-existing rock caught up in igneous rocks during their passage through the crust
ZEOLITES	a group of soft, light-coloured silicate minerals characterized by easy loss of water and sometimes by ion exchange capacity that makes them useable as water softeners
ZINC BLENDE	zinc sulphide, an important and decorative zinc ore
ZIRCON	zirconium silicate making tetragonal crystals